Energy
Matters

JACK HIGGINBOTHAM

Archway Publishing books may be ordered
through booksellers or by contacting:

Archway Publishing
1663 Liberty Drive
Bloomington, IN 47403
www.archwaypublishing.com
1 (888) 242-5904

ISBN: 978-1-4808-3398-2 (sc)
ISBN: 978-1-4808-3399-9 (e)

Library of Congress Control Number: 2016910875

Print information available on the last page.

Archway Publishing rev. date: 07/08/2016

Also by Jack Higginbotham is a textbook written for the Health Physics Society's professional summer school, *Applications of New Technology: External Dosimetry,* Medical Physics Publishing, 1996. Dr. Higginbotham has authored over 45 scientific papers and reports in the fields of radiation detection and measurement, electric demand forecasting, nuclear reactor reliability, nuclear accident analysis, environmental radiation transport and recently articles on aviation and space history. An admired educator, Professor Higginbotham has won awards for teaching and service excellence from students, Oregon State University and the Health Physics Society. During the middle decade of his academic career, Dr. Higginbotham answered his University's call to service as a senior university administrator. Beginning with his appointment as Associate Dean of the Oregon State Graduate School, then as Associate Vice President of Research, Dr. Higginbotham developed as an administrator specialized with assisting university offices undergoing leadership transition. His final university level administrative position was as President of the University Faculty Senate. An unwavering champion of academic shared governance, President Higginbotham successful convinced the University leadership to award all faculty a four percent raise (the largest in the previous ten years), to implement a smoke free campus policy, to streamline senate meeting procedures and to initiate a equity review process for salary levels of fixed term professional faculty positions.

Table of Contents

Preface

In the winter academic quarter of 2015, the Oregon State University Honors College conducted a class entitled Energy Matters. The course satisfied a University degree requirement that all students demonstrate the ability to synthesize information from various disciplines to foster critical thinking in a given content area. In the case of this class, the content area is science, technology and society. A secondary goal of the course was to give the Honors students a scholarly experience above and beyond the expectations for a traditional University class. During the early days of the term, the students showed themselves to be unusually gifted technical writers. This skill led to an expansion of the expectations for the class synthesis paper to include writing for a near future undergraduate audience. The goal was to bring their interests and importance ranking of Energy topics society is currently facing and why they matter to their generation.

The publication of their work in this format provided the students with an opportunity to experience the process of multiple authorship of a textbook and to provide a book reference to support their resumes as they apply to graduate schools.

About the Contributors

Brian Bemis

Brian is graduated from Illinois Valley High School, is currently studying electrical and computer engineering at Oregon State University and has an interest in material science and nanotechnology. He joined the solar car team as a freshman, is an Eagle Scout and past treasurer of the Lo La'am Geela Order of the Arrow lodge.

Hannah Bulovsky

Hannah is an undergraduate student in chemical engineering at Oregon State University focusing on environmental engineering. She is completing a thesis project investigating the use of zero-valence Iron nanoparticles for groundwater remediation. In her free time she enjoys reading hiking and cooking.

Jordyn Clarke

Jordyn Is an undergraduate student majoring in biochemistry and biophysics at Oregon State University. Outside of the sciences, she competes with the OSU intercollegiate horse show Association equestrian team.

Anthony Farr

Anthony is an undergraduate student at Oregon State University studying mechanical engineering. In high school he participated in first tech challenge robotics. He is pursuing a career in the field of mechanical design or renewable energy.

Pavel A. Grechanuk

Pavel is a motivated Nuclear Engineering student at Oregon State University. Some of his interests are space exploration, advanced nuclear reactor design and sustainable engineering practices. He is currently working as an analyst for the OSU Energy Efficiency Center where he drafts energy saving recommendations for industries in the region. In his free time, Pavel loves to spend time backpacking in the wilderness, traveling and snowboarding.

Jon Lawrence

Jon grew up in Oregon and he is a student at Oregon State University where he is majoring in chemical and environmental engineering. He has been a resident assistant for three years, most recently in Howley–Baxton Halls, home of the engineering learning community.

Musa A. Moussaoui

Musa is an undergraduate at Oregon State University studying Nuclear Engineering, Mr. Moussaoui is fascinated with all types of energy production, but the potential of nuclear power has piqued his interest. He is currently working on his undergraduate honors thesis on steam generators in nuclear power plants.

Michael Oatman

Michael is an undergraduate student at Oregon State University majoring in electrical engineering and minoring in computer science. He is a teaching assistant for an introductory engineering course and is also looking forward to applying for graduate schools. His brief experience and industry includes working on power supplies.

Acknowledgements

First and foremost I must acknowledge the efforts of Dr. Joe Zaworski, the originator of the class-Energy Matters. His professional integrity and passion for undergraduate education have been an inspiration to thousands of Oregon State University students.

To the student contributors, their lifelong pursuit of knowledge and understanding brought their academic skills to a point where they made the transition from students to intellectuals. The focused research to understand a topic through literature review, formulating ideas or models that explain the facts of the world around them, then produce a body of work to enlighten a specific future audience. I am fortunate to have been an observer of their transition and wish all the best to each as they move forward into the future.

A heartfelt thanks goes to Dr. Wayne Lei, of Portland General Electric Company, for presenting the class with insight into the current planning and investment of utilities. Also, thanks you to the staff of the Oregon Institute of Technology who took time to teach Oregon State students and to show the leading edge energy sources of the Klamath Falls campus.

Finally, to Catherine Lanier and Shirley Campbell a special thank you for giving students a safe, encouraging place to learn and for setting an example of a workplace focused on inclusion and success.

Introduction

By Jack Higginbotham

Energy Matters is the title of a course developed by Dr. Joe Zaworski, a course I took over when he retired. The overall objective was to provide university undergraduates with a venue to explore the role energy plays in the society of our world. Beginning with a review of basic science, thermodynamics, machine design, electrical/mechanical power and the progression of history as mankind uncovers new ways to move beyond his own muscle energy to change his environment.

There are alternate meanings to the phrase "Energy Matters". There is the declarative statement that communicates A fundamental societal concept – that while shelter, food and water are the basic elements of a persons survival, it is the mastery of Energy manipulation which gives a society not only control of these three essential elements in the now but also for the near future. A macabre example occurred during the decades between 1938 and 1958 when the world gained the scientific understanding of the conditions that must exist for a uranium 235 atom to split apart and convert a tiny fraction of its mass into raw energy. While Einstein's relationship of $E = mc^2$ was an outgrowth of his work on photoelectric effect in the first decade of the 20th century, the United States and many of its allies developed the technology to provide a sense of "shelter" for its citizens by serving as a retaliatory threat against those countries who wished to inflict physical harm on its population. This societal philosophy laid the foundation for International relations that have permeated the world ever since.

Another illustration of an application of the phrase "Energy Matters" was expressed during my own undergraduate academic career at Kansas State University in the spring semester of 1977. Four years previously a number of the international oil generating states decided to restrict production resulting in a global shortage of gasoline and fuel oil. The impact on my life was it took longer for me to earn enough money working for my local school district to fill my car's gas tank and it took more time for me to wait in line to have that opportunity. An understanding of why "Energy Matters" came from the professor teaching macroeconomics class. His position was that society should take a long view of the future and purchase every drop of oil under the sand the Middle East. I have been sufficiently schooled in the burgeoning expectation of political correctness to know that such a

1

statement was ethnically offensive. His following point was the value of goods and services produced by fossil fuels over the next two centuries would far outweigh the near-term energy value of such a finite resource and thus it is in the economic interest of our society to husband our own fossil resources for future utilization instead of squandering them with shortsighted efforts of keeping gasoline prices low. Therefore if other societies were going to sell their resource to us, we should purchase every drop. I took his position in a different direction. I begin to wonder why so much of the United States population depends upon use of fuel oil for residential and commercial heating? I spent my formative years as an Air Force brat living on base housing where natural gas was the primary heat source but it was during my early collegiate experience that I became aware of the energy history of the Northeast United States - where Americans moved from using timber as a heat source, to coal, then oil. Why? A question I have found sufficiently interesting to define by life for the past three decades. Another reason "Energy Matters" – it is just plain interesting!

That professor died of pancreatic cancer the next semester. I do not remember his name but I do remember his idea and the effect it had on my thinking and understanding. An understanding much different than what he was trying to convey because I brought my own set of experiences and priorities to the question of energy.

In the context of teaching the Honors College class "Energy Matters" at Oregon State University during the winter quarter of 2015, I engaged the students to set the priority for subjects to covered in the class and to what depth while facilitating an environment of collaborative investigation where each students became a subject expert who communicated to the rest via a writing project. Such an effort would not follow the path of those previous "research papers" written for a grade but which would be addressed to a future audience – the following year's class of "Energy Matters".

What follows is a textbook, grouping nine chapters written by student subject experts. The list of subjects was generated during a free flow, in-class discussion and each student chose the subject of their chapter. One third of the students were funded to make "field trips" to deepen their understanding of the subject. After weeks of research, chapter drafts were shared with the rest of the class and through this peer-review process; each author received eight independent reviews of their work. The process also gave each student a directed reading

2

assignment of the other eight topics for the class. The chapter titles provide the stage, but the thesis of each chapter is wholly the work to the student author. The chapters are: Life cycle assessment, Coal, Biomass, Nuclear reactors, Solar, Geothermal energy, Hydropower, Natural gas and Wind energy.

Chapter 1. Life Cycle Assessment
By Richard Soteros

Life cycle assessment (LCA)[1] is a tool used by experts from various regulating organizations, including the International Organization for Standardization and the Environmental Protection Agency in the United States. LCA measures the environmental performance of products and processes and can be used as a tool to investigate ways to minimize effects on the environment.

The assessment takes a cradle-to-grave approach[2] where data is collected throughout the entire lifespan of a product. It begins with the gathering of raw materials to create the product and ends when all materials are returned back into the earth. LCA is an estimation of the cumulative impact throughout all states of the product's life. It does this by assembling an account of relevant energy and material inputs in combination with environmental releases that are based on potential environmental impacts due to involved inputs and outputs. The reason a life cycle assessment is preferred over other methods of impact study is because it includes areas of impact that other analyses leave out. This provides a more accurate assessment of the environmental trade-offs involved in creating a product or process.

Life cycle assessment analyses all five of the phases of a products life. These phases are raw material acquisition, manufacturing, use, maintenance, and final disposal. The first phase, raw material acquisition, takes into account all of the emissions and extraction impacts caused by the removal of raw materials from the earth. For example, gathering of wood used in the creation of a house would be accounted for. The transportation of raw materials is also a factor during this phase. Second, the manufacturing stage is where most other analyses start. This phase includes the transformation of the raw materials into usable products. With the example of wood, this is the stage where the raw wood would be turned into lumber for use in a house, chair, or other wood products. The use phase includes the part of a products life that consumers influence. This is when a house would be

---

[1] Life Cycle Assessment (LCA) Chapter 4. (2014, August 5). Retrieved March 16, 2015.

[2] Scientific Applications International Corporation (SAIC). (n.d.). Life Cycle Assessment: Principles and Practice. 68-C02-067.

lived in or a chair would be sat upon. The analysis of this phase would include the energy consumption that the house uses for heating, the water pumped in and out of the house, and the waste that is produced and pumped out of the house. The maintenance stage takes that a bit further by including things like a new roof for a house in 20 to 50 years. This stage can vary drastically depending on the product or service being analyzed. The last phase, final disposal, is the grave in the cradle-to-grave assessment. When a consumer disposes of a product, those materials in the product are analyzed for their impact over the time they will be decaying when back in the earth. The impact could be low, like products made from organic material, or very high, as is with nuclear waste and some types of plastics.

There are four components to how the life cycle assessment is created. The researcher, or research group, begins by defining the goal of the study. This is simply for terminology and focus, not so much for guiding the study in a particular direction. This goal also defines the scope of the assessment by establishing a clear context and setting boundaries on what inputs and outputs will be included. After collecting those inputs and outputs, the researcher(s) analyze the inventory taken. In this component they will quantify the inventory. Examples of this typically include items such as kWh of energy demanded, gallons of water used, volume of raw materials extracted, etc. In order to make units comparable, a characterization matrix is created. The equation for this can be seen as

$$h = Qg \qquad (1.1)$$

where g is the energy flows from the inventory list, Q is the characterization values, and h is the standard vector[3] . An example of this equation could look somewhat like the following, as used to determine the impact of greenhouse emissions:

$$g = \begin{bmatrix} 1.0 \ kg \ CO_2 \ emissions \ to \ the \ air \\ 0.5 \ kg \ CH_4 \ emissions \ to \ the \ air \\ 0.1 \ kg \ NH_3 \ emissions \ to \ the \ air \end{bmatrix}$$

---

[3] Von der Assen, N. (2014, May 7). How can I find the mathematical model/formula of life cycle impact categories? Retrieved March 11, 2015.

$$Q = \begin{bmatrix} 1 & 25 & 0 \ (\textit{climate change in kg } CO_2) \\ 0 & 0 & 2.45 \ (\textit{terrestrial acidification in kg } SO_2) \\ 0 & 0 & 0.092 \ (\textit{marine eutrophication in kg N}) \end{bmatrix}$$

$$\therefore$$

$$h = \begin{bmatrix} 1(1) + 0.5(25) + 0.1(0) \\ 1(1) + 0.5(0) + 0.1(2.45) \\ 1(0) + 0.5(0) + 0.1(0.092) \end{bmatrix}$$

$$\Downarrow$$

$$h = \begin{bmatrix} 13.5 & kg \ CO_2 \ (\textit{climate change}) \\ 0.245 & kg \ SO_2 \ (\textit{terrestrial acidification}) \\ 0.0092 & kg \ N \ (\textit{marine eutrophication}) \end{bmatrix}$$

Once the inputs and outputs are quantified, the impact can be assessed. The results from this component give a number value to the overall human and environmental effects. Typically only a single product or service is being analyzed so a simple addition of the h values will determine impact. If the data is being expanded to show total impact over an industry, the equation can be changed to multiply by quantity. An example of this could be when calculating the total global warming potential (GWP) of the previously calculated greenhouse emissions[1]. For CO_2, the GWP equation would be:

$$GWP = (3.5kg \ CO_2) \times (\textit{quantity of this product being produced})$$

Last, interpretation, ties back into the previous three components. The final resulting impact may rely heavily on the scope decided at the beginning of the assessment. Likewise, quantifying abstract notions of impact, many of which could be indirect or obscure, takes a great deal of interpretation by the researcher(s) and could lead to a very different conclusion. For this reason, it is important to understand who conducted the assessment when using the results.

There are many benefits of conducting a life cycle assessment for products and services. By performing a LCA, decision makers can also analyze the trade-offs in a quantified fashion that takes the environment into account. It helps them evaluate the environmental consequences of their business procedures and could, in the end, help them reduce their environmental footprint by seeing their impacts quantified. If desired, it can be used to select the least impactful product, process, or service. This can be used for positive media or, if the firm is environmentally focused, help a firm move closer to its overall mission. Occasionally the least impactful product can also be more cost effective and perform better. If a product or service uses fewer inputs to produce its end result, then it typically costs less and thereby reduces a company's bottom line. Another benefit is that LCA provides a way to compare different business opportunities. These opportunities could be between the creation and sale of different products, or it could be between the first use and recycled use of a single product. Performing a LCA comparison between different uses can identify shifts in impact that happen between the different life cycles. This can cut costs for some businesses if the benefits of reusing a product are significant. Finally, since LCAs delve so deeply into the total impact of a product or process, the impacts can be seen on local, regional, or global scales. For example, if a LCA is performed on a nuclear reactor the LCA could measure the impact on the local community neighboring the reactor and could also measure the impact of the outputs in the region.

There are certain limitations of conducting and using the life cycle assessment method. Conducting a LCA is very resource and time intensive. When deciding if a LCA is cost effective, the decision makers need to weigh the availability of the data needed, the time needed to conduct the assessment, and the financial resources required. Although having a LCA for every product and for use in every decision would be ideal, the process is costly and may not be a priority if the business does not have an interest in environmental impact. For businesses where performance is key in the choice to produce a product, reducing performance to reduce environmental impact may not be a viable option. Since this type of assessment will not show which unit performs better and can only potentially show a difference in cost effectiveness, it is important to be used as only a component of the full analysis. For this reason, the LCA method may be more useful to larger corporations or on the governmental level, or for those who aim to make life cycle management a priority. Those who aim to manage the life cycle of a product or policy toward being more sustainable, across every phase of the cycle, would find this type of assessment helpful. As mentioned before the interpretation phase influences all parts of the total analysis, but it also extends to the use of the analysis itself. This phase could include judgment calls made by managers or a panel of experts. In either case, life cycle assessments are simply one tool to be used to influence business decisions or policy formation.

Braishali Dutta and Vijaya Raghavan[4] from McGill University in Montreal, Canada provide an example of a life cycle assessment. Their analysis focused on pyrolysis biochar systems to determine the greenhouse gas balance, carbon cycling, and the economics of producing biochar from various agricultural wastes and residues . Their goal for the research was to provide insight into alternative biochar sources other than typical bioenergy crops and the resulting energy and environmental impact. This included sources such as forest residues and corn fodder. They found that these two sources both had a reduction of greenhouse gas emission compared to traditional biochar. Even in the same study, the researchers found corn fodder to have a further reduced greenhouse gas effect than forest residue, indicating that corn fodder may have a greater economic potential. This is a great example of how a LCA can provide information when deciding between two alternatives. The study also found, however, that even with decreased greenhouse gas emissions these sources were better means for carbon soil sequestration than as a fuel source. Even though the two alternatives studied may have reduced total emissions, the opportunity costs associated with them may cause them to be more costly than typical biochar sources. So although the life cycle assessment showed it favorable in environmental impact, the total viability of the sources may not be any better than the current option.

ISO Standards

The International Organization for Standardization [5] (ISO) develops and publishes international standards for various items including materials, products, processes and services. These standards can be in the form of requirements, specifications, guidelines or characteristics. The mission of the ISO, and the purpose of standardization, is to ensure safe, reliable, and high quality products for consumers. Most items that receive an ISO standard are also good for the environment. By standardizing consumer goods, ISO helps to minimize waste and error in production. In general, this increases productivity and efficiency in the industry.

---

[4] Dutta, B., & Raghavan, V. (n.d.). A Life Cycle Assessment of Environmental and Economic Balance of Biochar Systems in Quebec. Int J Energy Environ Eng, 5.106. Retrieved February 16, 2015.

[5] International Organization for Standardization. (n.d.). Retrieved February 16, 2015, from http://www.iso.org/iso/home.htm

Development of a standard through ISO takes six steps between when a need for a standard is introduced to when it is published. When ISO gets a request for a standard to be created, a group of experts at ISO write a proposal. This proposal includes the reason why the standard is needed. Once the proposal is accepted, a working draft of the proposed standard is created. The draft written at this time may differ drastically from the finalized version. The draft is shared with a technical committee, which consists of experts in the industry that the standard would affect, and with the ISO's Central Secretariat. If a consensus is reached in the technical committee, then the draft is given to all of ISO's national members. The national members are asked to give comments on the draft, either in favor of or against. If a consensus can again be reached, then the draft is finalized. The finalized version gets sent to all members of the ISO across the world. These members have the right to vote on the implementation of the international standard. If a majority is reached, the final draft becomes an official ISO International Standard.

Throughout this process, ISO conforms to certain principles. These principles are used in the creation of all ISO standards since their implementation. First, ISO only creates standards as a response to a need in the market. The ISO only responds to requests and does not seek out opportunities by which to make standards. These requests can come from individuals, corporations, regulatory agencies, or governing bodies. Second, details of the standards are based on the aggregate of global expert opinion. The concept of the Technical Committee is used to gather those expert opinions into certain subcommittees. Committees vary greatly based on subject, leading to the formation of 268 existing committees to date. Because these committees can include such specialized fields, the scope of the standard, the definition of its application, and its content are all evaluated to a great extent before being given to any ISO members. The last principle lays its roots in democracy. A consensus is needed from four out of the six steps of the development process. The proposed standard will not move to the next step in the process until consensus is reached, and will surely not be implemented if a majority is not reached from the final vote of all ISO members. This system forces drafts of standards to be amended until all parties are content with the finalized draft that will eventually be implemented internationally.

The benefits of the ISO International Standards can be seen across three dimensions, including business, society, and government. In the business dimension, there are five aspects that are typically improved. The first aspect is monetary savings. Adhering to standards can optimize operations and improve a company's bottom line. Standardizing methods by which companies produce or provide their product can streamline their operating procedures and reduce overall cost. This reduction of expenditures helps raise a company's bottom line in most cases. The second aspect is enhanced customer satisfaction. Standards allow a consumer to expect a certain quality of the product or service they are purchasing. If consumers can know what to expect when they buy, they are more likely to be satisfied and become return customers. Third is access to new markets. Once an international standard is implemented countries require products that adhere to that standard, usually no higher or lower. This can help prevent trade barriers since the international community can simply use the ISO standard for its quality requirement, and since producers should be producing to at least that standard. The fourth aspect is an increase in market share. Companies that provide products and services that comply with ISO standards can use compliance as a selling point. This increases competitive advantage to a certain extent, depending on the market environment. The fifth aspect is an increase in environmental benefits. Since ISO standards typically account for environmental impact, compliance helps reduce negative externalities on the environment and others. For certain industries this can be a large factor in their personal image.

In society, ISO standards affect most industries. Public services, like roads, libraries, and education, have standards that they must meet for the general wellbeing. Private products, like children's toys or electronics, have standards mostly due to safety. Standards pertaining to the manufacturing of toys help keep harmful chemicals away from children. Medical products, like packaging on instruments, also have standards to ensure safe use and guarantee cleanliness. Consumers of these ISO products can be confident of their quality and safety. As previously mentioned the ISO takes into account the environmental impact of products when developing standards, but the ISO also develops standards for the environment itself. The ISO has created international standards on air, water, and soil quality. Some of these standards include emissions, such as those from radiation that is produced by nuclear power plants, coal energy plants, car exhaust, and burned jet fuel.

For the government, ISO's standards are used in the formation of public policies. Because they are international standards, governments typically wish to comply with them. The ability to enforce standards is minimal on the international level. The ISO will take legal action against individuals and companies who chose to not comply, but enforcement is difficult if a nation as a whole disregards international standards. Governments who do comply generally prefer the use of ISO standards. Since experts develop the standards they are easily adopted and trusted, although noncompliance sometimes stems from mistrust of these experts. Another benefit to governments comes from the simplicity of translating a standard. When using an international standard in a policy, that policy is easily understood by other nations. Policies regarding international trade prefer the use of standards because production standards in one nation can naturally meet the consumption standards of another nation, and vice versa.

When producers or suppliers adhere to an ISO standard, they can apply their product or service for standard certification. Being certified adds credibility to claims made about the quality of the product or service. It can mean that they meet a certain prospect that the consumer is expecting. Certification can even be legally required in some industries. The actual certification process is not done by the ISO, however. Accredited third party bodies manage the certification of products and services, allowing the ISO to strictly develop and publish standards. An official certification for a product or service would read *ISO 9001:2008 certified*, for example.

The ISO also provides other standard corporate services, including management systems and audits. The first is a model for standardizing management of a company. This is typically used when setting up a system but is also applicable to its operation as well. This model was developed in the same fashion all other ISO standards are developed. Since the model is specific to management, it can be applied in any industry. The claimed benefit of following this model management system includes: more efficient use of resources, improved risk management, and increased reliability of products to customers. The audit service exists as a check to see how well a company is meeting its goals. This is mostly an internal service but can also be used to check that a company is adhering to the standards needed for certification.

11

EPA Standards

Along with standards posed by the ISO, the Environmental Protection Agency (EPA), a federal agency in the United States, strives to protect human health and the environment through the development of regulatory policies and guidelines that are legally enforceable. Although the ISO takes the environment into consideration when forming its standards, it is more focused on consumer protection. The EPA, like the name suggests, is primarily focused on protecting the environment, but also indirectly protects consumers in the process. The EPA is unique from other agencies because it can write policy by itself. These policies target manufacturers, private enterprises, utilities, air zones, national wildlife regions, and many other sources of human interaction with the environment. Furthermore, if these policies are not followed, the EPA press criminal charges against offenders.

In 1970, the legislation passed by the U.S. Congress that created the Environmental Protection Agency in 1970, granted the Agency the authority to promulgate regulations to carry out its mandate to protect the environment and granted enforcement authority to enable compliance with its policies. The Agency also creates guidance and implementation documents to lead individuals and corporations towards compliance. The EPA maintains a list of significant guidance documents, as called for by the Office of Management and Budget.

The process[6] by which the EPA poses regulations is much like that of the ISO, although somewhat less stringent. The process starts by proposing a regulation, either initiated by the EPA or by a request received from Congress, interest groups, or individuals. The EPA then researches the issue and writes a draft for a potential future regulation. The draft is posted in the Federal Register so the public can review it and submit comments. These drafts are rewritten based on the input received and subsequent drafts are posted. Once the final draft is written, the regulation is then submitted to the Federal Register and codified in the Code of Federal Regulations. At that point in the process, the regulation starts to take effect, unless the regulation specifies otherwise.

Enforcement[7] is an integral part of the EPA's mission. It will

---

[6] The Basics of the Regulatory Process. (n.d.). Retrieved February 16, 2015, from http://www2.epa.gov/laws-regulations/basics-regulatory-process

[7] Enforcement Basic Information. (n.d.). Retrieved February 16, 2015, from http://www2.epa.gov/enforcement/enforcement-basic-information

take civil or criminal enforcement if noncompliance occurs. Since the EPA is specifically environmentally focused, its ability to enforce regulations cannot extrapolate past environmental statues or regulations, either federal or state. However, the language granting the EPA explicit enforcement authority is fairly vague so it can sometimes enforce regulations that have only small overlap with environmental issues. Most of these issues pertain to human health. Its specific goal is *Environmental Justice*,[8] which can only be achieved when all people are protected equally from environmental and health hazards, while having a say in the decision-making process of their living environment. This, more than anything else, drives their decision whether to prosecute potential offenders. Generally, if the EPA files a civil case they are trying to bring polluters back into compliance and may charge them a fee, although relatively small in relation to profits. If a criminal case is filed, however, it means the EPA is trying to prosecute those who deliberately violated the law. In criminal cases, violators may face incarceration or correctional fees. The fees charged in criminal cases are typically far greater than those in civil cases. The third option that exists is cleanup enforcement. In these cases, the EPA finds the agent responsible for the contamination, and then works with them to organize a cleanup strategy. The EPA may require that the polluter clean up the contamination themselves, or will sometimes allow them to contract out to a cleanup crew. This method of enforcement is used mostly for instantaneous events, but is sometimes used for long-term pollution.

---

[8] Environmental Justice. (n.d.). Retrieved February 16, 2015, from http://www.epa.gov/environmentaljustice/

Problems:

1) Given *g* and *Q*, find *h*.

$$g = \begin{bmatrix} 3.5 \ kg \ CO2 \ emissions \ to \ air; \\ 1.5 \ kg \ CH4 \ emissions \ to \ air; \\ 0.4 \ kg \ NH3 \ emissions \ to \ air \end{bmatrix}$$

$$Q = \begin{matrix} 1, 25, 0 \ (climate \ change \ in \ kg \ CO2 - eq); \\ 0, 0, 2.45 \ (terrestrial \ acidification \ in \ kg \ SO2 - eq); \\ 0, 0, 0.092 \ (marine \ eutrophication \ in \ kg \ N - eq) \end{matrix}$$

Solution:

$$h = \begin{bmatrix} 3.5 \times 1 + 1.5 \times 25 + 0.4 \times 0 \\ 3.5 \times 0 + 1.5 \times 0 + 0.4 \times 2.45 \\ 3.5 \times 0 + 1.5 \times 0 + 0.4 \times 0.092 \end{bmatrix}$$

$$\Downarrow$$

$$h = \begin{bmatrix} 41 \ kg \ CO2 - eq \ (climate \ change) \\ 0.98 \ in \ kg \ SO2 - eq \ (terrestrial \ acidification) \\ 0.0368 \ in \ kg \ N - eq \ (marine \ eutrophication) \end{bmatrix}$$

2) Assuming the following value for the h matrix for a for a single process, what would total impact be if this process were to be repeated every other month for a year?

$$h = \begin{bmatrix} 7.2 \ kg \ CO2 - eq \ (climate \ change) \\ 0.58 \ in \ kg \ SO2 - eq \ (terrestrial \ acidification) \\ 0.0163 \ in \ kg \ N - eq \ (marine \ eutrophication) \end{bmatrix}$$

Solution:

$$h = \begin{bmatrix} 43.2 \ kg \ CO2 - eq \ (climate \ change) \\ 3.48 \ in \ kg \ SO2 - eq \ (terrestrial \ acidification) \\ 0.0978 \ in \ kg \ N - eq \ (marine \ eutrophication) \end{bmatrix}$$

3) Given *h* and *Q*, find *g*.

$$h = \begin{bmatrix} 20 \ kg \ CO2 - eq \ (climate \ change) \\ 0.8575 \ in \ kg \ SO2 - eq \ (terrestrial \ acidification) \\ 0.0322 \ in \ kg \ N - eq \ (marine \ eutrophication) \end{bmatrix}$$

$$Q = \begin{bmatrix} 0 & 25 & 0 \ (climate\ change\ in\ kg\ CO2-eq) \\ 0 & 0 & 2.45 \ (terrestrial\ acidification\ in\ kg\ SO2-eq) \\ 0 & 0 & 0.092 \ (marine\ eutrophication\ in\ kg\ N-eq \end{bmatrix}$$

Solution:

$$g = \begin{bmatrix} 0 + 20 \div 25 + 0 \\ 0 + 0 + 0.8575 \div 2.45 \\ 0 + 0 + 0.0322 \div 0.092 \end{bmatrix}$$

$$\Downarrow$$

$$g = \begin{bmatrix} 0.8 \ kg\ CO2\ emissions\ to\ air \\ 0.35 \ kg\ CH4\ emissions\ to\ air \\ 0.35 \ kg\ NH3\ emissions\ to\ air \end{bmatrix}$$

4) What is each product's impact if the total impact of 13 products is given as

$$h = \begin{bmatrix} 64.8 \ kg\ CO2-eq \ (climate\ change) \\ 5.44 \ in\ kg\ SO2-eq \ (terrestrial\ acidification) \\ 0.039 \ in\ kg\ N-eq \ (marine\ eutrophication) \end{bmatrix}$$

Solution:

$$h = \begin{bmatrix} 4.98 \ kg\ CO2-eq \ (climate\ change) \\ 0.418 \ in\ kg\ SO2-eq \ (terrestrial\ acidification) \\ 0.003 \ in\ kg\ N-eq \ (marine\ eutrophication) \end{bmatrix}$$

5) What is the global environmental impact of a product that is used in 2.2 million homes globally if an individual product is expressed as:

$$h = \begin{bmatrix} 0.002 \ kg\ CO2-eq \ (climate\ change) \\ 0.0005 \ in\ kg\ SO2-eq \ (terrestrial\ acidification) \\ 0.00006 \ in\ kg\ N-eq \ (marine\ eutrophication) \end{bmatrix}$$

Solution:

$$h = \begin{bmatrix} 4400 \ kg\ CO2-eq \ (climate\ change); \\ 1100 \ in\ kg\ SO2-eq \ (terrestrial\ acidification) \\ 132 \ in\ kg\ N-eq \ (marine\ eutrophication) \end{bmatrix}$$

Chapter 2. Coal
By Michael Oatman

As time moves forward and nations develop, so does the demand for electricity to power society. Coal is the most utilized resource for electrical power in developing countries such as China, India, and Russia because it is plentiful, reliable, and cost effective source of energy. Coal power is still heavily relied on in the U.S. as well because of its cost and quantity. However, the reliance on coal as an energy source has also come with a heavy price on the global environment, as well as the local society around the coal mining and burning areas. If mankind still wants to utilize coal for its electric power potential, changes need to be made in order to protect the global environment and the people in coal mining areas.

Understanding the Technology

The life cycle of coal comes in three stages: mining, refining, and burning. Each of these stages includes different technologies and has different waste products as a result. It is important to understand each stage of the coal energy cycle to see how coal fits into the global energy picture.

The mining stage is split into two categories: surface mining and underground mining. Surface mining consists of any form of mining where the coal vein is located close to the surface and the topsoil is removed by either bulldozers or explosives. Since the coal is close to the surface, setting up an underground mining facility would be economically inefficient. Surface mining also allows for a higher recovery rate of the coal seam compared to underground methods, obtaining up to 90% of the coal that can be recovered. In the U.S., 66% of the total coal produced is surface mined.[9] Surface mining techniques include area mining, open pit mining, contour mining, auger mining, and mountain top removal, whereas the underground methods consist of long wall mining or room and pillar mining.

Area mining is a technique used in relatively flat areas, primarily in the west coast and Midwest regions where the geography contains plains and rolling hills. The technique consists of cutting a series of parallel rectangular strips into the land known box cuts, which may be

---

[9] "Coal Mining Methods". Energy and Minerals Field Institute at the Colorado School of Mines

hundreds of yards wide and over a mile long. As a box cut is being made, the waste from making the cut, which contains all the top soil, vegetation, and rocks sitting on top of the coal, is placed into the previous box cut that has already been stripped of the coal inside. Once the previous box cut is filled with all of the waste, it is covered with more topsoil and reseeded so the vegetation that was once there can grow back.[9]

Open pit mining is a technique implemented for very large seams of coal that can be up to 100 feet thick. Since the seams of coal are extremely thick, a large amount of land needs to be removed to get to the coal. The waste removed from the pit is either placed on adjacent land or even shipped to another part of the open pit where the coal has already been extracted and needs to be backfilled. Open pit mining is a large-scale operation where production levels can exceed 10 million tons of coal per year.[10]

Contour mining and Auger mining normally go hand in hand together, and take place in really steep mountainsides where the coal seams are outcropped and visible from the surface. For contour mining, a cut is made on the slope where the coal is located and, once the overburden is removed and placed aside, the coal is removed. A series of cuts are made along the mountainside where the coal seam exists until the coal resources are depleted.

Auger mining is utilized when the height or steepness of the mountainside renders the removal of the overburden uneconomical to the miners, and instead of cutting and removing the overburden, a large mining auger—essentially a large drill bit—is used to create a large bore directly into the coal seam and the coal is removed.

The final surface mining technique used is mountain top removal. Mountain top removal is primarily seen in the Appalachian Mountain Range, especially in states such as West Virginia and Kentucky. In order to reach the coal seam at the center of the mountain, holes are strategically drilled around the mountain and filled with dynamite. Once the dynamite is detonated, the mountaintop overburden is loosened and removed and the coal seam is exposed. This allows for the entire seam of coal to be collected. Although mountain top removal

---

[10] Mark Squillace, "The Environmental Effects of Strip Mining", *The Strip Mining Handbook* (1990)

requires much more engineering, capital, and skill than the previously listed methods, it is a popular method because of its efficiency.

Long wall mining is an underground mining method that is widely used across the world and is starting to grow in popularity in the U.S. It consists of cutting out a coal bed—averaging 800 feet wide, 7,000 feet long, and 7 feet tall—underground and using an automatic shearer to travel from one end to the other over and over until the coal is extracted. Once the coal is sheared off of the coal bed, it is placed onto a conveyor belt that transports the coal out of the mine. The shearer is also attached to hydraulic roof supports, that way the roof of the mine doesn't collapse when the shearer removes the coal holding it up, and these supports move along with the shearer so they advance forward when the shearer advances forward. The roof of the mine is allowed to collapse behind the hydraulic supports. Long wall mining offers a couple advantages because it is an automated operation that continues until the entire coal bed is removed, meaning less manpower is needed, and it has a recovery rate that averages around 57% nationwide. The disadvantages to using long wall mining include the capital cost of the equipment, the complexity of the system, and the need for well-maintained ventilation inside the mine to take care of the dust and the methane released.

Room and pillar mining is an underground mining process where the caves are organized into rooms and pillars. Miners first drill into the coal, pack the holes with explosives detonate the explosives, and then extract the coal from the mine. This process creates a room where the operation can then continue further. The pillars are blocks of coal purposely left untouched during the mining process, typically 40 to 80 feet in width, and are used to support the roof of the cave and keep it from collapsing. Since coal is left behind in order to keep the roof from collapsing, room and pillar mining is typically less efficient than long wall mining. In order to recover even more coal once the mine's coal resources are exhausted; mining companies may try to remove the coal pillars starting from the deepest pillars moving outwards. This phase in the room and pillar mining process is called "retreat mining".

Coal Preparation
Once the coal is removed from the mine, it is transported to a preparation center before being sent to the coal-fired plants for burning. The crude coal that is shipped from mines varies widely in size, ash

content, moisture content, and sulfur content. For example, crude coal mined in the U.S. can range anywhere from 3% to 60% ash.[11] Today, over one third of the total coal mined in the U.S. is sent to coal preparation facilities before being sent to the power plants in order to prepare the coal to meet market specifications. The purpose of refining and cleaning the coal before burning it is to remove impurities from the coal, since most coal removed from mines can contain up to 60% impurities that consist of shale, slate, clay, or other materials that cannot be burned.[12] By removing the materials that cannot be burned, the coal's heating value is increased, producing more thermal energy, making the power plants more efficient. Ash levels are also reduced by removing the other materials mixed with the coal because they will not end up at the bottom of the boiler as ash. Another benefit to removing the impurities from the coal before burning it is reducing the toxic emissions such as sulfur dioxide (SO_2) and nitric oxides (NO_X).

In the preparation process, the coal is first crushed into smaller pieces and screened as either coarse or fine. Subsequent processing steps depend on feed stream particles size. For fine particles—ranging from 0.2 mm to 1 mm—the entire coal bed is submerged into water, where the smaller, carbon rich pieces of coal float to the surface and the heavier particles containing more impurities fall to the bottom. The top layers of the submerged coal bed undergo a dewatering process to remove all of the moisture from the coal.

Impurities are segregated from the coarse particles by forming a solid solution of coarse coal and finely pulverized magnetite. The magnetite additive provides a denser solution than achieved with water suspensions and hence higher specific gravity materials can be separated. The coal/magnetite solution is run through a cyclone separator where lighter coal particles are separated from the heavier contaminant streams. Coal is then rinsed and the magnetite removed with magnetic separators.[12]

Removing the moisture in the coal is essential to raising the heating value of the coal. Any moisture in the coal can also cause freezing problems if the coal is being shipped to colder regions, as well

---

[11] "Coal Preparation Plants". *Inspector's Guidance Manual*. Stationary Sources Branch and Air Pollution Control Division in Colorado, www.colorado.gov
[12] "Coal Preparation", *Meeting Projected Coal Production Demands in the U.S.A.* Virginia Tech

as add weight to the coal which making it more expensive to ship. The dewatering process consists of a series of screens, thickeners, and centrifuges, which removes most of the water inside the coal. Thermal dryers utilize hot gases to remove the rest of the moisture in the coal.[13]

The most prominent waste from a coal preparation plant is black water - a sludge consisting of a mix of water and very fine particles of coal at are extremely difficult to separate. This mixture is discarded into man-made impoundments.

Coal Fired Plants

There are multiple types of coal-fired plants that utilize different techniques to transfer the heat from the coal fire into electrical energy. Each technique has different advantages and disadvantages regarding efficiency, waste products and cost effectiveness. The types of coal fire plants include pulverized coal combustion (PCC), cyclone coal combustion, fluidized-bed combustion, and integrated gasification combined cycle (IGCC). Differences between the coal-fired plants lie largely in the process of heating up the coal and maintaining the coal fire. Once the coal fire is running, the transfer of heat energy to electrical energy is generally the same in all of the different coal fire plants.[14]

Coal Burning Methods

The pulverized coal combustion, PCC, plant is used in 92% of coal-fired plants in the United States, because it is the most robust way of providing power on a large scale while still meeting environmental and economic standards. In a PCC plant, the coal is ground up into a very fine powder with the same consistency of talcum powder; meaning at least 70% of the powder will pass through a 200-mesh sieve (0.074 mm sieve size).[14] The purpose of grinding the coal up into a fine powder is it increases the surface area between the coal and the oxygen for combustion to occur, optimizing the chemical reaction and creating a much hotter fire. Burning the fine coal powder in a PCC plant is almost as efficient as burning a gas. Since coal in powder form burns almost instantaneously once inside the boiler, changing the flow rate of coal

---

[13] "Coal Cleaning", United States Environmental Protection Agency, www.epa.gov
[14] "Available and Emerging Technologies For Reducing Greenhouse Gas Emissions From Coal-Fired Electric Generating Units", United States Environmental Protection Agency, October 2010, www.epa.gov

powder entering the boiler can control heat generation inside the boiler. If the coal were to be in bigger chunks and burned for a while, it would be much harder to cool the fire down because stopping the flow of coal into the furnace wouldn't mean the fire inside the furnace would be put out. A fine powder also means starting the fire up is much easier because it is easier to ignite the coal powder than it is to ignite bigger chunks. Once the coal is pulverized, it is blown into the boiler with hot air and combusted. If we assume that the coal is made up of purely carbon and has negligible amounts of other elements such as hydrogen, oxygen, nitrogen, and sulfur, then the chemical reaction for the combustion of coal is:

$$C + O_2 \rightarrow CO_2$$

$$(2.1)$$

Example Problem: Given 24 grams of coal and an unlimited supply of oxygen, how many moles of carbon dioxide are produced? Assume the coal contains a negligible amount of elements in it other than carbon.

Example Problem Solution: 2 moles of CO_2. From the reaction equation of Eq. (2. 1), one mole of carbon yields one mole of CO_2.

This chemical reaction is heavily simplified because the chemical makeup of coal varies widely depending on the type of coal, the heat the coal is burned at, and the technique used to burn the coal. PCC plants are split into three categories that characterize the heat and pressure of the steam they produce.

The most common type of PCC is the subcritical pulverized coal unit (SubCPC), which typically produces steam at 2,400 psi and 1,000 °F, and has a heat energy to electrical energy efficiency of roughly 34.3%.[15] SubCPC units also produce about 22,800 kg/h of ash that needs to be removed from the boiler. The other two types of PCC plants are the supercritical pulverized coal (SCPC) and the ultra-supercritical pulverized coal (USCPC) units. Although much more expensive than the SubCPC, these two types of PCC plants have been growing in popularity now that technology is advanced enough to make them reliable, as well as for their higher energy transfer efficiencies. The SCPC plant produces steam at roughly 3,530 psi and 1,050 °F, and has an electrical efficiency ranging from 37% to 40%. The USCPC plant produces steam at 4,640 psi and 1,110 °F, with electrical efficiencies ranging from 44% to 46%.

---

[15] "The Future of Coal," Massachusetts Institute of Technology, 2007, Pages 19-22.

Not only do these two types of PCC plants operate at higher efficiency, but they also use less feed air and coal per hour compared to the SubCPC plants to output the same amount of power. Typical values of fuel and oxygen feed rates for a 500 MW$_e$ plant are a function of plant type. SCPC and USCPC plants use 1,940,000 kilograms of feed air per hour and 164,000 kilograms of coal per hour, compared to 2,450,000 kilograms of feed air per hour and 208,000 kg of coal per hour in the SubCPC plants for the same power output.

The downside of the PCC plants is they need emissions controls on the output of the boiler to clean up the toxic flue gases that harm the environment, adding capital and maintenance costs to the plant as a whole. Fortunately, these emissions control systems can remove up to 99% of the sulfur fumes and 90% of the NO$_X$ gases in the flue gas, but by removing these toxins from the air emissions, it produces wet solids that are combined with the slurry as waste. A standard SubCPC plant will produce around 41,000 kg/h of wet solids as waste that needs to be removed from the site.[15]

Another way to burn coal is through cyclone coal combustion. Cyclone coal combustion boilers were designed to burn lower grade coal that has higher moisture or ash contents, and accomplishes this by strategically placing burners inside the boiler to create high temperature flames that rotate in a cyclonic pattern.[14] Before entering the boiler, the coal is pulverized into a powder of 4-mesh size and carried into the boiler with the primary air stream. Once the coal enters the cylindrically shaped boiler, a secondary air stream jets along the outside creating a swirl within the boiler. Due to centripetal force, the larger pieces of coal will be pushed to the outside of the swirl, which keeps them circulating in the vortex until they finish burning. A tertiary jet of air is injected into the middle of the cyclone in order to control the vortex size and heat.[16] The molten ash from the burned coal creates a slag that forms at the bottom of the boiler and is drained through a tap opening. Cyclone coal combustion boilers make up roughly 6% of all the coal-fired plants in the U.S., and operate at subcritical steam levels, producing 35% to 38% electrical efficiency. The major drawback for cyclone coal combustion is the high NO$_X$ emission rates compared to other methods.

---

[16] "Cyclone Fired Wet Bottom Boilers", IEA Clean Coal Centre, United Kingdom, www.iea-coal.org.uk

A much newer technology that is utilized to burn lower grade coals, much like the cyclone coal combustion boiler, is the fluidized-bed combustion boiler. The fluidized-bed combustion boiler gets it names because it suspends all of the solid fuels on upward jets of air during combustion, which creates a turbulent mixture of gas and solids that acts like a bubbling liquid. The solid fuels used inside the boiler for combustion is a combination of both coal and a sorbent, typically limestone or dolomite, which can absorb up to 95% of the sulfur pollutants released by the combustion process. Fluidized-bed combustion also avoids high NO_X emission because it burns fuel at low temperatures around 1,500 °F, compared to 2,450-2,750 °F in a PCC. This avoids the formation of NO_X gases because NO_X typically starts forming around 2,500 °F. [17] The disadvantages to fluidized-bed combustion are that it doesn't fix the CO_2 emission at all, it has similar waste problems the rest of the plants run into, and the tube on the inside of the boiler can start eroding due to the turbulent fluid running inside of it. Because fluidized-bed combustion systems are relatively new, they only consist of 2% of the total coal-fired plants in the U.S.

The last method for coal-fired plants in the U.S. is integrated gasification combination cycle (IGCC). IGCC methods have yet to be implemented on a large scale in the U.S. because it is a very new technology. In the IGCC the coal is converted into a synthetic gas by taking a coal and water mixture and reacting it with oxygen at high temperature and pressure to produce a combustible gas that is composed of carbon monoxide and hydrogen. The reaction for coal gasification is:

$$C_xH_y + \frac{x}{2}O_2 \rightarrow xCO + \frac{y}{2}H_2 \qquad (2.\,2)$$

In this equation, the variables x and y are the chemical makeup of the coal being gasified. Once the coal is gasified, a water-gas-shift reaction is used on the syngas to purify the H_2 fuel even further. The chemical reaction for the water-gas-shift reaction is:

$$CO + H_2O \rightarrow CO_2 + H_2 \qquad (2.\,3)$$

---

[17] "Fluidized Bed Technology", Office of Fossil Energy, www.energy.gov

After looking at the two chemical reaction equations above[18], it is easy to see how the slurry of water and coal is reacted with oxygen to make a gas for burning. Once the fuel is made, it is transported to the boiler and burned to create steam for the electric generation cycle just like the other types of coal-fired plants. The electric efficiency from an IGCC plant is 38.4%, very comparable to a PCC operating at supercritical conditions. IGCC plants need more air fed into the system, roughly 2,880,000 kg/hr, than a PCC does because it needs to provide for both the gasification and the combustion cycles.[15] IGCC plants are more advantageous than other methods of burning coal because they can produce high efficiencies, fuel flexibility, less need for post combustion emission control systems—most toxins can be removed in the development of the syngas, less solid waste, and less water consumption.[14] Another huge development in IGCC technology that makes it favorable for the future is since the CO_2 is created during the gasification stage, the CO_2 is already under high pressure making it easier to capture and store it before it is released to the environment. The disadvantages to the IGCC plant are design complexity, high construction cost, and poor performance at high altitudes where natural air pressure is much lower than at sea level.

Heat to Electrical Energy Conversion

All coal-fired power plants have the same basic concepts, but use different methods to achieve the end goal of transforming heat energy into electrical power for the grid. If we want to describe the coal-fired plant as a thermodynamic system, it needs to have a hot body, a heat engine, and a cold body. In the case of the coal plant, the coal fire is the hot body. Once the coal fire is running hot in the boiler, whether it is from PCC, cyclone combustion, fluidized-bed combustion, or IGCC, water is pumped to the boiler through tubes where it is heated into steam. This steam is held at both high temperature and pressure, and it is then routed through a turbine—the heat engine in the thermodynamic system, where both the temperature and pressure gradient from one end of the turbine to the other causes the steam to move rapidly through the turbine, causing the blades to spin at extremely rapid rates and creating AC electrical power, which is then routed from the power plant to the

---

[18] "Considerations For IGCC Power Plants Designs", Yisha Mao, Stanford University, 2012.

power grid and supplied to the public. After the steam passes through the turbine, it is piped to a heat sink, such as a lake or river, which acts as the cold body in the thermodynamic system. Once heat is transferred from the steam to the cold body, it will condense back into water, where it is then pumped back to the boiler for reuse and the cycle continues.

The Economics Behind Coal

In the U.S., energy demands nationwide are on an enormous scale, and so is the need for a resource to supply a large chunk of those energy needs affordably. Coal supplied 39% of the electrical power used in the U.S. in 2013, compared to 27% from natural gas, 19% from nuclear, and 13% from renewable resources.[19] It supplies the most power in America because of its abundance and cost effectiveness.

The U.S. Energy Information Administration (EIA) believes that as of January 1, 2013, the amount of recoverable coal in the existing mines in the U.S. sits at 18.7 billion short tons. On top of that, the EIA also estimates that the total reserves of coal, or the demonstrated reserve base, in the U.S. is at 481.4 billion short tons. After taking into account current mining technologies and other limiting factors, the EIA estimates that the total amount of recoverable coal in the U.S. is 257.6 billion short tons, which also accounts for 27.3% of the estimated global world energy reserves. The U.S. holds a very large portion of the total coal in the world. If coal production rate grows by its projected 0.3% per year from 2012 to 2040, the U.S. has enough coal to last 180 years if no new reserves are found.[20] The numbers speak for themselves; there is an absolute abundance of coal in the United States meaning it is an easy source to rely on to deliver large amounts of power.

The second reason coal is relied so heavily upon in the U.S. is it is extremely cheap. As of 2013, the national average price for coal mined in the U.S. and sold to the electric power sector—electric utilities and independent power producers—was $45.21 per short ton of coal.[21] Example Problem: How much does it cost a subcritical pulverized coal plant to make a single kWh of energy? Remember that a subcritical

---

[19] U.S. Energy Information Administration, June 13, 2014, www.eia.gov
[20] "How Much Coal is Left", U.S. Energy Information Administration, 2014, www.eia.gov
[21] "Annual Coal Report 2013", U.S. Energy Information Administration. Table 34, www.eia.gov

pulverized coal plant uses 208,000 kg/hr of coal to produce 500 MW$_e$. Assume the plant purchased the short ton at the national average price. Example Problem Solution: 2.07¢. Use dimensional analysis to solve. Start with the given data about subcritical pulverized coal combustion plants.

$$\frac{208,000 \text{ kg}}{\text{hr}} \rightarrow 500 \text{ MW}_e$$

Convert input to system into short tons and output of system to MWh:

$$h * \frac{208,000 \text{ kg}}{\text{hr}} * \frac{\text{short ton}}{907.185 \text{ kg}} \rightarrow 500 \text{ MW}_e * \text{hr}$$

See how much energy 1 short ton will get you, knowing that a short ton sells for $45.21:

$$1 \text{ short ton} * \frac{45.21 \text{ dollars}}{\text{short ton}} \rightarrow \frac{500 \text{ MW}_e\text{h} * 907.185}{208,000}$$

Calculate how much energy 1 dollar into the system will produce, convert to kWh and find the cost per kWh:

$$1 \text{ dollar} \rightarrow \frac{500 * 907.185}{208,000 * 45.21} \text{ MWh} * \frac{1000 \text{ kWh}}{\text{MWh}} = \frac{2.07 \text{ ¢}}{\text{kWh}}$$

The technology used today is advanced enough that by the time the coal is mined, processed, and burned, the electrical power can be sold to the end consumer in the U.S. for anywhere from 6 - 15¢/kWh, depending on location and the government policies, for a national average of 8.6¢/kWh.[22] According to the EIA, the total national average for the price of electricity in all sectors—residential, commercial, industrial, and transportation—from all sources of energy combined is 9.84¢/kWh.[23] The cost of electrical power produced by coal is an entire cent cheaper than the average price of power, which makes it a much more desirable option.

---

[22] Richard A. Muller, *Energy for Future Presidents*, 2012, Pages 273-279.
[23] "Electricity Power Annual", U.S. Energy Information Administration, 2012, Table 2.4

Societal Impacts of Coal

Although coal may be an abundant, easy to harvest electrical energy form and economically cheap, it imposes a heavy toll on local communities in coal mining, processing, and burning areas. Many of these societies are heavily impacted across all aspects of the community, whether it's economic, political, or the general health of the people.

West Virginia is the second largest coal producing state in the nation, producing 157,994 thousand short tons of coal per year, and has the highest number of employees in the coal industry out of any state at 20,454 employees.[24] It also ranks third lowest among all states in the U.S. for personal income per capita at $34,477 in 2012.[25] It doesn't help that there is a tradition of West Virginia being a coal mining state, and there is a very popular notion that education isn't needed when anyone can easily obtain a coal-mining job. According to the U.S. census in 2007, West Virginia ranks last among the states with 17.3% of the population that is 25 or older to have a Bachelor's degree or higher.[26] Even though a person can make a decent living off of coal mining, coal-mining jobs are normally not long term. If there is ever a downturn in the number of coal mining jobs available, the community cannot find alternative means of generating income and end up suffering the economic consequences. According to the U.S. National Center for Biotechnology Information, Appalachian communities that had the heaviest coal mining, had the lowest socioeconomic conditions, and that these communities also typically had low economic diversification, employment in professional services, and educational attainment rates. The coal industry may not be the only factor that plays into these statistics, but it is impossible not to acknowledge that it plays a major role in these problems. Considering that the coal industry is the second largest employer in the state behind Walmart shows that, although the coal mining industry does provide jobs, the wages are so low that it leaves the community around it with very little income.[27]

---

[24] "The West Virginia Coal Economy 2008", West Virginia University, 2010, Pages 12 and 20

[25] "Per Capita Personal Income by State", Bureau of Business and Economic Research, UNM, 2013

[26] "Persons 25 Years Old and Over with a Bachelor's Degree or More, 2007", U.S. Census, www.census.gov

[27] Omar Ghabra, "How the Coal Industry Impoverishes West Virginia", The Nation, 2014, www.thenation.com

Pollutants from coal can also be responsible for health problems within the surrounding community and affect all major body organ systems. It contributes to heart disease, cancer, stroke, and chronic lower respiratory disease, which are four of the five leading causes of mortality in the U.S.[28] A study performed on the Hunter Region in Australia, a heavily mined area, found that coal mining areas have health affects across all demographics in the society. Adults have higher mortality rates due to lung cancer, various chronic heart, respiratory, and kidney diseases, as well as higher hospitalization rates for chronic obstructive pulmonary disease (COPD) and hypertension. In the study, they also looked at the Appalachian regions of the U.S. and compared mountaintop coal mining areas, non-mountaintop mining areas, and non-mining areas. They found that both of the mining areas had significantly higher mortality rates dues to cardiovascular diseases than the non-mining areas, with the mountaintop coal mining areas having the highest rates. They also found that mining areas have higher percentages of children born with birth defects, neural tube defects, and low birth weight, as well as higher amount of miscarriages and stillbirths. An example of the affect mining has on the neurological development of children can be seen in Heshun, Shanxi Province in China, which is responsible for producing 25% of China's coal and 5.6% of the world's. Heshun has one of the highest rates in the world for neural tube defects among children.[29]

The health and quality of life impacts on the surrounding community caused by the coal industry also reaches beyond diseases and sickness. Communities in the vicinity of mountaintop removal sites are a prime example of the physical damage the coal industry can have on the people in the area. Runoff from the rainfall picks up all of the dirty remnants from the blast and contaminates the local water supplies. Gregory J. Pond, environmental biologist for the EPA, found that 90% of 27 Appalachian streams tested that existed below valley fill sites for mountaintop removal mining failed the Clean Water Act standards, whereas all 10 of the streams located in non-mining areas passed.[30] Not

---

[28] "Coals Assault on Human Health", Physicians for Social Responsibility, 2009.
[29] "Health and Social Harms of Coal Mining in Local Communities: Spotlight on the Hunter Region", Beyond Zero Emissions, Melbourne, October 2012.
[30] David C. Holzman, "Mountaintop Removal Mining: Digging into Community Health Concerns", National Center for Biotechnology Information, 2011.

only does mountaintop removal contaminate water supplies, but also the lack of topsoil and trees on the recently removed mountaintops can cause rainwater to flow all the way down the mountainside and collect in dangerous amounts at the bottom of the valleys, typically where the people live, creating a flash flood.[31] The flash floods created by the removal of mountaintops ravage local communities and can be responsible for thousands of dollars in damages to homeowners in the area. Finally, mountaintop removal also emits a significant amount of dust into the air, lowering the air quality in the vicinity and creating respiratory problems for the local community, which also adds to the higher rates of mortality due to the cardiovascular diseases discussed earlier.

Environmental Problems Caused By Coal

Using coal as a source for electrical energy creates problems within the environment in every aspect of the coal life cycle, from mining, to preparation, and, finally, from burning. The most prominent environmental impacts that coal casts onto the surrounding ecosystem are the physical damage, waste products, and emissions.

Coal mining is harmful to the ecosystems that it takes place in because of the massive removal of chunks of land to get to the coal. This is especially seen in surface mining techniques such as area mining, open pit mining, and mountaintop removal. In 1977, the Surface Mining Control and Reclamation Act was passed as an attempt to limit the physical effects the coal mining process has on the land, by mandating that coal companies restore the land to its original shape after the coal is mined. However, the law is poorly enforced and reclamation efforts are difficult to accomplish - especially for mountaintop removal. With these large-scale operations of surface mining removing large amounts of land, the ecosystem and animal populations that existed on that land are either displaced or destroyed. Since these mining operations remove the topsoil that contains all the nutrients for growth, the land left behind is barren, and it is extremely difficult for the ecosystem to recover on the reclaimed land.[32] Another problem these large-scale mining sites run into is that rainfall falls onto the piles of waste coal, absorbs toxic

---

[31] "Coal Controversy in Appalachia", NASA Earth Observatory.
[32] David A. Buehler, Katie Percy, "Coal Mining and Wildlife in the Eastern United States: A Literature Review", University of Tennessee, 2012.

chemicals from the coal, and flows as runoff into the groundwater.[33] This problem is countered by burning the waste coal on site, which instead releases CO_2 emissions as well as SO_2 and NO_X pollutants since the coal hasn't been refined yet.[34]

The primary waste product that comes from coal is sludge, also referred to as slurry or blackwater, which is a mixture of coal, ash, water, and any other solid waste or chemicals that come from coal. The sludge is normally placed into large man made facilities or pumped into underground mines that have been exhausted of their resources. There is little regulation governing the building of sludge impoundments. Sludge impoundments are usually built at the bottom of a valley next to the preparation or power plant, and are not completely sealed from the surrounding environment, meaning the heavy metals and toxic waste can easily seep into the surroundings.[35] Another issue that arises with these impoundments is the risk of a spill. The biggest slurry spill to date is the Buffalo Creek Disaster in 1972, where a local slurry pond in Logan County, West Virginia broke open and spilled 120 million gallons of water and 35 million cubic feet of waste material, killing 123 people and leaving numerous more injured or homeless.[36] Another big slurry spill occurred in 1996 in Buchanan County, Virginia, when an underground mine impoundment leaked its contents through the abandoned mine to the other side of the mountain, where the slurry then flowed into the Big Sandy River. The spill is estimated to have peaked at 1000 gallons-per-minute, and effects of the spill were observed up to 35 miles downstream, killing an estimated 21,000 fish.[35]

The last big environmental issue that is raised when talking about coal is the emissions. Coal releases a number of pollutants to the environment when it is burned, the biggest being SO_2, NO_X, and CO_2. In the U.S., the burning of coal is responsible for 40% of the hazardous air pollutants. SO_2 and NO_X gases released from the burning of coal are

[33] Martha Keating, "Cradle to Grave: The Environmental Impacts from Coal", Clean Air Task Force, 2001.

[34] Phillip J. Lloyd, "Coal Mining and the Environment", Energy Research Institute, University of Cape Town.

[35] "Accidental Releases of Slurry and Water from Coal Impoundments Through Abandoned Underground Coal Mines", Mine Safety and Health Administration, Pittsburgh Safety and Health Technology Center.

[36] "Environmental Justice Case Study: Buffalo Creek Disaster", University of Michigan.

largely responsible for rising acidity in the environment, and when these gases mix with water they create acid rain.[37] Acid rain ravages the forests and wooded areas in the environment, as well as raises the acidity in freshwater sources. The CO_2 emissions from coal are enormous, and these levels cannot be avoided if coal is to be relied on as an energy source since the fundamental chemical reaction from the burning of coal creates CO_2. Current CO_2 levels in the atmosphere are at 390 ppm; pre-industrial CO_2 levels were 280 ppm. The rise in CO_2 levels since the industrial revolution are widely believed and proven to be the leading cause behind the global temperature increase of 0.64°C over the last 50 years. Global climate change is the biggest and scariest issue the world is facing regarding the environmental damage coal is causing, because the severity of the repercussions are unknown by scientists.[22]

My Personal Experience with the Coal Industry

When I was a senior in high school, I had the fortune of going on a weeklong immersion trip to the Appalachian region with some of my classmates. The purpose of the trip was to see the societal and environmental impacts the coal industry was having on the surrounding area. What I saw in the communities of West Virginia that we visited was completely unexpected, and gave me a new perspective not only on my own personal place in the world, but also on the means a society will go through to acquire cheap affordable energy. The most memorable aspect of the trip for me was the scenery as we drove through the countryside from rural town to rural town, and seeing the conditions most of the people in West Virginia lived in. Most of the roads we drove on that were separate from the highways were severely cracked and were in bad need of being repaved. We passed through numerous housing communities that were strictly made up of mobile homes parked in a neighborly fashion in the middle of grassy clearings, without running water or electricity. All the houses we saw were small, old, and very beat up just like the towns they were in. Our state guide, born and raised in West Virginia, told us that the two biggest employers in the state were Walmart and the coal industry, and that just about anyone and everyone in the state either worked, or knew someone who worked, in the coal mines at one point or another in their lifetime. She had family

---

[37] "Emissions of Hazardous Air Pollutants from Coal-fired Power Plants", Environmental Health and Engineering, Inc., 2011.

members who were coal miners and were suffering symptoms of black lung, which, according to her, was still a very relevant and problematic sickness within the communities in West Virginia. She also told us that high school graduation rates were extremely low because people will live their whole lives believing that they don't need education as long as the coal industry needs miners. My amazement of the impacts caused by the coal industry was propelled even further when we drove to a mountaintop removal site. On the way up to the site we passed by several little ponds on the side of the road that were literally orange, which our guide explained was just the runoff water from the rain that collected mining waste. Once at the mountaintop removal site, it was really incredible to see such a massive bare piece of land in the middle of a heavily wooded and green mountain range. The man who gave tours around the site was an advocate at the volunteer organization ilovemountains.org, and had his mobile home parked next to the site where the rest of his family resided. We noticed his mobile home had bullet holes in it, and he explained that some local coal miners shot up his home for his stance on mountaintop removal. This amazed me because in my mind the negative effects of the coal industry clearly outweigh the benefits of it, especially to the local community, but that story showed me that some of the people who live there and have been a part of the coal industry their whole life don't see the negative impacts as much as they see the jobs it provides.

Regarding the Future of Coal

If the U.S. wants to continue to utilize coal as a major power provider in the future, it needs to address some of the societal and environmental effects coal has. First and foremost, the U.S. needs to stop the use of mountaintop removal as a way of surface mining, due to its sheer destruction of the ecosystem and the health effects it imposes on the surrounding community. Also, the coal industry needs to find a way to cut back on the production of solid wastes in the form of sludge. The best way to cut back on the solid waste production is to find a way to better filter out the fine coal particles in the slurry and burn them.[38] There also needs to be more research done on underground injection of slurry and other waste from the coal process, because theoretically, the

---

[38] "Coal Waste Impoundments: Risks, Responses, and Alternatives", National Academy Press, Page 169.

waste will be deep enough underground that it will be below the local water table, and therefore will not contaminate the groundwater supply while also keeping it away from ecosystems on the surface. In a study done on underground injection, none of the water sources around the injection showed signs of contamination.[39] In order to face the ever looming problem of climate change and rising CO_2 levels, IGCC technology needs to implemented more, due to its high energy output and lower carbon output, as well as the ability to perform carbon capture and storage on it is much easier than in the standard PCC plants.

[39] "An Evaluation of the Underground Injection of Coal Slurry in West Virginia", West Virginia Department of Environmental Protection, Page 33.

Problems:

1. If you are running an IGCC plant and have 10 moles of oxygen to react with a very large supply of coal and water, how many moles of hydrogen gas can you make from the water-gas-shift reaction?

   Answer: 20 (10 moles of oxygen gives 20 moles of carbon monoxide in the gasification stage, which in turn gives 20 moles of hydrogen gas in the water-gas-shift reaction)

2. Suppose you are managing a project for an underground coal mine that utilizes longwall mining. If your mine costs a total of 1 million dollars for construction, maintenance, and operation, and contains 50,000 short tons of coal, how much money do you expect to make once the coal resources are exhausted? Assume your mine operates at the national average for efficiency, and sells the coal to the customer for the national average price.
   Answer: \$288,485 (using 57% mining recovery of coal, \$45.21 per short ton, subtract the initial cost of the mine)

3. New emissions regulations are released and you are forced to install emissions control systems in your 500 MW_e subcritical PCC plant that will remove 99% of the SO_2 and 90% of the NO_X in the emissions from the burning process. If the coal you are burning releases emissions that are 82% CO_2, 13% SO_2, and 5% NO_X, how many kilograms of SO_2 and NO_X are being released into the environment per hour? A subcritical PCC emits 2,770,000 kg/hr of stack gas.
   Answer: 3,601 kg/hr SO_2, and 1,385 kg/hr of NO_X.

4. You are looking at your electricity bill and you see that in the last month, you used 500kWh of energy. How many kilograms of coal were used to help supply power to your home? Assume all coal energy comes from subcritical PCC plants at the national average partition of all electricity generated in 2013. Round to nearest whole kilogram
   Answer: 81 kilograms (39% of total energy from coal, you know 208,000 kg of coal are need for 500 MW_eh in SubCPC conditions, set up a simple ratio to find amount needed for your energy consumption).

5. You are designing an IGCC power plant that is required to supply 500MW_e. If you know the average heat content of a short

34

ton of coal is 20.21 million Btu's, how many pounds of coal are needed per hour to meet specifications? Round to 2 significant digits.

Answer: 440,000 pounds per hour (convert Btu's/hr into kW for a single short ton of coal, use efficiency of IGCC and specification to find the number of short tons per hour needed and convert into pounds).

Chapter 3. Biomass

By Jon Laurence

Biomass is the largest source of renewable energy in the United States[40]. All biomass energy comes from plants and is the product of photosynthesis, which is the utilization of electromagnetic radiation from the sun to convert carbon dioxide to carbohydrates. This includes biomass sources such as human waste, which ultimately comes from the energy stored in plants. Humans have relied on biomass for longer than any other source of energy. In 2013, renewable energy accounted for 10% of the energy consumed in the United States and the mix of energy sources are: Wood (22%), biofuels (22%), biomass waste (5%) and other biomass (50%)[41].

Wood has been used for thousands of years for heating and cooking. Theoretically, biomass is a carbon neutral source of energy if harvest rates match growth rates. Many current policies aim to increase biomass energy production based on this idea. New research has shown that biomass is not always sustainable and the perfect balance between biomass production and utilization is hard to achieve. This chapter will explore biomass by focusing on woody biomass, conventional ethanol production from corn, and methane production at landfills and municipal wastewater treatment plants.

Woody Biomass

Oregon consistently leads the United States in timber production with 3,327 million board feet of timber harvested in 2010. A board foot is a measure of volume and is equivalent to 1 ft. by 1 ft. by 1 in. The Oregon Department of Energy offers tax credits, energy loans, and other incentives for utilizing woody biomass for energy. There are some worries that if strong reliance on woody biomass for energy production lead to negative impacts on the forest ecosystem such as habitat loss and soil quality depletion.

---

[40] Oregon Department of Energy. Oregon's Bioenergy Resource Map. [Cited 28 February 2015] Available From
http://www.oregon.gov/energy/RENEW/Biomass/Pages/Bioenergy_map.aspx
[41] United States Energy Information System. Renewables. [Cited March 1 2015] Available From http://www.eia.gov/kids/energy.cfm?page=renewable_home-basics

The primary method energy from woody biomass is utilized is through combustion. Wood [42] contains many different complex hydrocarbons bound together with lignin. In general the composition of wood is 50% carbon, 44% oxygen, 6% hydrogen, the rest is inorganic material. During the combustion process about half of the wood's original mass is converted into water. The other half is converted into carbon dioxide while producing about 8,600 Btu/pound, or 20 MJ/kg. When wood is heated above 500° C it undergoes pyrolysis, which is the thermochemical decomposition of organic materials into gas in the absence of oxygen. Primary combustion is the burning of solid material directly, such as the charcoal formed by pyrolysis. Secondary combustion is the burning of gases liberated by pyrolysis. This represents 85% of the mass and 60% of the heating value. Complete combustion is the conversion of organic material to carbon dioxide and water. Incomplete combustion products include carbon monoxide and many different hydrocarbons. Incomplete combustion is what leads to particulate matter, and other forms of air pollution. Combustion is an oxidation-reduction reaction where reduced carbon in fuel is oxidized to carbon dioxide and O_2 is reduced[43]. To explore the combustion of wood Nazaroff, et. al.,[43] looked at Douglas fir. The effective chemical formula for Douglas fir is $CH_{1.45}O_{0.18}N_{0.002}$ with 0.8% of the mass as ash. With stoichiometric amount of air, the combustion reaction for Douglas fir is

$$CH_{1.45}O_{0.580}N_{0.002} + 1.07(O_2 + 3.78N_2) \rightarrow CO_2 + 0.725H_2O + 4.06N_2$$

Assuming air is 21% O_2 and 79% N_2 and all nitrogen products are in the form of N_2.

To approach ideal combustion, air is needed in excess, typically at 50-100% more than the stoichiometric amount. In reality, complete combustion is never achieved and the products contain carbon monoxide, nitrous oxides, and various unreacted hydrocarbons.

---

[42] United States Environmental Protection Agency. Burn Wise Workshop. Wood Combustion Basics. [Cited 28 February 2015] Available From http://www.epa.gov/burnwise/workshop2011/WoodCombustion-Curkeet.pdf
[43] Nazaroff W, Alvarez-Cohen L. Environmental Engineering Science. New York (NY). John Wiley and Sons, Inc. 2001. p. 358-362.

To convert biomass into electrical energy, the Rankine cycle is often used. The Rankine cycle is an idealized vapor power system that can be used to explore the major concepts found in a modern power plant[44]. The working fluid in practice is water; throughout the cycle the water exists in four different states separated by four different processes. Heat energy is transferred to the working fluid in the boiler: the hot body. Heat has to be transferred to the environment to return the working fluid to its original state: the cold body.

The first law of thermodynamics states that energy can be changed from one form to another, but the total amount of energy in the universe is constant. We can use this fact to help us understand the Rankine Cycle. The Rankine cycle is a steady state process; to simplify our analysis we will neglect the kinetic and potential energy contributions of a fluids location and velocity. The following equation is a statement of the steady state first law of thermodynamics. Equation (3. 1) states the rate of energy input into the system has to equal the rate of energy leaving the system, because we have chosen a steady state process.

$$0 = \sum_{in} \dot{m}_{in} \, \hat{h}_{in} - \sum_{out} \dot{m}_{out} \, \hat{h}_{out} + \dot{Q} + \dot{W}_s \qquad (3.1)$$

Where $\hat{h}$ is specific enthalpy, $\dot{m}$ is the mass flow rate, $\dot{Q}$ is the rate of heat transferred into or out of the system, and $\dot{W}_s$ is the rate work is done by or on the system. Two notes about this equation: when $\dot{W}_s$ is negative, it represents work being done by the system and the time rate of change of work and energy is equal to power. Enthalpy is a convenient collection of energy terms defined as the sum of internal energy and flow work. Values of enthalpy can be calculated many ways, but values for water at various temperatures and pressures can be found in steam tables, making calculations with water relatively simple.

At state 1 water is a superheated vapor. This vapor is expanded through a turbine where it also cools. The change in enthalpy of the water through the turbine is equal to the work produced by the turbine. The vapor does not condense completely through the turbine, because liquid water is corrosive and can damage the turbine. The steam exits the turbine at state 2 and enters a condenser. The condenser is at

[44] Koretsky M. Engineering and Chemical Thermodynamics. Hoboken (NJ). John Wiley and Sons Inc. p. 164-168.

constant pressure so heat needs to be removed to condense the steam into saturated water. The heat transport out of the steam can only occur in the presence of a temperature differential so a cold body is needed as a driving force for temperature change. The amount of heat energy transfer out of the condenser is equal to the change in enthalpy of the water from state 2 to state 3. The saturated liquid water is then sent through a compressor where the pressure is increased but the volume ideally remains unchanged. Because the volume is approximately constant through this process it requires a relatively small amount of work, compared to the work produced by the expansion through the turbine. The high-pressure liquid exits the compressor at state 4 where it goes to the boiler. Heat from combustion is transferred to the water until the water exits as a super-heated vapor. The combustion chamber of the boiler acts as the hot body and the amount of heat energy transported into the fluid is equal to the change in enthalpy of the water through the boiler. Super-heated steam exits the boiler at state 1. The efficiency of the Rankine Cycle, η, is the ratio of work provided divided by the heat supplied; in this case it is the power produced by the turbine minus the power to operate the compressor divided by the heat absorbed from the boiler.

$$\eta = \frac{\left|\dot{W}_{Turbine}\right| - \dot{W}_{Compressor}}{\dot{Q}_{input\ from\ boiler}} \qquad (3.\,2)$$

Example Problem

In a typical paper mill producing process steam, steam enters a turbine at 100 bar and 500° C and exits at 10 bar. How much power per kg of steam does this process produce? Assume an ideal reversible Rankine Cycle.

A reversible process is one that has no change in entropy; it is the limiting case where a process takes an infinite amount of time to happen. The system remains in equilibrium and there is no energy transfer to the surroundings. The values of $\hat{s}_1$, specific entropy; $\hat{h}_1$, specific enthalpy; $\hat{v}_1$, specific volume; and $\hat{w}_1$, specific work, are be used with a steam table to find enthalpy and entropy of steam at 100 bar and 500° C in this case:

$$\hat{h}_1 = 3{,}373.6\frac{kJ}{kg} \quad and \quad \hat{s}_1 = 6.5965\frac{kJ}{kg}.$$

The specific values of state parameters can be found following a Rankine flow path. The pressure drops to 10 bar through the turbine, but the entropy remains constant. The value of enthalpy that corresponds to this state can be found by equating $\hat{s}_1$ and $\hat{s}_2$ then using a steam table to find that $\hat{h}_2 = 2{,}782.7 \frac{kJ}{kg}$. The steam enters the condenser and exits at state 3 in a 10 bar saturated liquid state. The steam table provides the value: $\hat{h}_3 = 762.79 \frac{kJ}{kg}$. To find the enthalpy of state four the following assumption is made. The work used by the pump to compress the liquid water can be found with the following equation assuming the volume of liquid water does not change as a function of pressure.

$$\widehat{W}_{pump} = \hat{h}_4 - \hat{h}_3 = \hat{V}_3(P_4 - P_3) \tag{3.3}$$

Therefore:

$$\hat{h}_4 = \hat{h}_3 + \hat{V}_3(P_4 - P_3) = 762.79 \frac{kJ}{kg} + 0.001127 \frac{kg}{m^3}(10{,}000 kPa -$$
$$1{,}000 kPa) = 772.9 kJ/kg$$

-- the power per kg of steam produced by the process.

The total work can be found by adding the work consumed by the compressor and the work provided by the turbine using the equation:

$$\widehat{W}_{total} = \widehat{W}_{pump} + \widehat{W}_{turbine} \tag{3.4}$$
$$\widehat{W}_{total} = (\hat{h}_4 - \hat{h}_3) + (\hat{h}_2 - \hat{h}_1) \tag{3.5}$$
$$\widehat{W}_{total} = \left(772.9 \frac{kJ}{kg} - 762.79 \frac{kJ}{kg}\right)$$
$$+ \left(2782.7 \frac{kJ}{kg} - 3373.6 \frac{kJ}{kg}\right)$$
$$\widehat{W}_{total} = -580.79 kJ/kg$$

The efficiency of the cycle can be found using by using Eq. (3.2).

$$\eta = \frac{W_{turbine} - W_{compressor}}{\hat{h}_1 - \hat{h}_4}$$
$$\eta = \frac{590.9 - 10.1}{2600.7}$$

$$\eta = 22\%$$

Wood pellet fuels are made from the residual sawdust from sawmills and at times directly from forest biomass. The sawdust is dried, compressed into pellets, and sold for both commercial and residential use. The final product is very consistent compared to other sources of wood heating fuel such as cordwood. The amount of energy available in cordwood depends on the species of tree and the moisture content. A cord is the standard unit used to measure the volume of firewood. It is defined as the volume of wood that can be stacked in a 4 ft. by 4 ft. by 8 ft. volume. The actual volume of wood depends on the size of the wood pieces and how it was stacked.

Bear Mountain Forest Products is a manufacturer of wood pellets in Oregon. Eric Laurence, the plant manager was interviewed about their process. They make 85,000 tons of pellets each year at their mill in Brownsville solely from sawmill residuals. The process starts by drying sawdust from moisture content of 50% to 5%. Water is trapped in the wood structure so it takes more time and energy for the sawdust to dry as compared to an equivalent amount of water in an open environment. All of the energy for the drying process comes from dried sawdust and a small amount of pellets. This represents about 20% of total incoming sawdust, which is equivalent to 70 million Btu per hour of drying capacity. The dryers used have the ability to use both sawdust and natural gas. As of the beginning of 2015 sawdust is used to operate the drying process. The cost to operate the drying process with sawdust and natural gas are within 1-1.5% with sawdust slightly less expensive.

Dry sawdust is combusted completely in a burner and the exhaust gasses provide the heat for drying. The combustion products along with the sawdust to be dried are pulled through the dryer by an induced draft from two, 300 horsepower fans. The sawdust and gas travel through 230 feet of ducts where the sawdust continues to dry. The sawdust is separated from the exhaust gas by a cyclone.

The sawdust exits the cyclone and goes to a 300 horsepower hammer mill where it is broken into smaller particles. This process happens after the drying process because it takes more energy to "cut" wet sawdust than to "grind" dry sawdust. The sawdust is sized and sent to one of 4 pellet mills where it is compressed into wood pellets. The dust is evenly distributed into a pelletizing chamber. The pelletizing chamber consists of rollers inside a die, which is a circular ring with pellet-sized holes. The die rotates at about 200 rpm and is powered by four 400 hp motors. The sawdust is compressed to 22,000 psi, pushing the sawdust through the die and creating pellets. This process reduces the moisture from 5.5% to 2.5%. The pellets remain bound due to the lignin in the wood. The pellets come out of the mill at 300° F and are sent to a cooler where they are cooled to the ambient temperature with one 125 hp. fan. A small fraction of pellets and all of the fine material that was not pelletized are sent to the burner where they are used as fuel.

The exhaust gas, which contains the products of wood combustion, is sent to a scrubber where the gas is saturated with water. A cyclone removes the majority of the liquid water, which is recycled back to the scrubber. The saturated exhaust gas then travels through a wet electrostatic precipitator. An electrostatic precipitator works by charging, collecting, and removing particles[45]. A large magnitude voltage is set up between two electrodes; at Bear Mountain the potential difference is 50,000 V. This large voltage produces a corona discharge near the negatively charged electrode. A corona discharge is an electrical discharge brought about by the ionization of gas molecules. Free electrons from the ionized gas are repelled from the negative electrode and collide with molecules in the process gas. The negatively charged molecules are attracted to the positively charged electrode and collide with particles in the process gas. Particles continue to collect negatively charged ions and become increasingly attracted to the positive electrode where they are deposited. In a wet electrostatic precipitator, water is used to flush the particles off of the electrode periodically. Two, 150 hp. fans pull the gas though the electrostatic precipitator and discharge it through a stack.

When operating at steady state, the process uses electrical energy at a rate of 2.1 MW. Along with the equipment energy use discussed, there are about one hundred 3-5 hp motors used for various jobs such as conveyance. This energy use represents about 3% of the energy available in the pellets that are produced. The problem at the end of the chapter explores how much energy is used in the drying process.

In the United States 80 percent of the paper made is by the Kraft process[46]. The paper industry uses 2,200 trillion Btu per year, equivalent

[45] Neundorfer. Electrostatic Precipitator Operation. [Cited 28 February 2015] Available From https://www.neundorfer.com/FileUploads/CMSFiles/ESP%20Operation[0].pdf
[46] Office of Air Quality Standards U.S. Environmental Protection Agency. Available and Emerging Technologies for Reducing Greenhouse Gas Emissions from the Pulp

to 14% of the energy consumption of the manufacturing sector in the United States. The Kraft process uses wood chips and two main chemicals, sodium hydroxide (NaOH) and sodium sulfide (Na_2S), to make pulp[47]. Around the globe 130 million tons of Kraft pulp are produced each year. This represents two thirds of total pulp production and 90% of chemical pulp. The Kraft process is able to achieve 97% chemical recovery of process chemicals. A byproduct of the Kraft process is black liquor, which contains lignin residues and the inorganic process chemicals. Black liquor is used as a biomass source for energy production and provides 64% of the pulp industries energy needs.

The Kraft process starts by sending wood chips in to a continuous digester where the chips are introduced to NaOH and Na_2S under high temperature and pressure[48]. The chips are then blown out of the digester; the sudden decrease in pressure turns the mixture into a brown fluffy substance. This mixture contains pulp and black liquor that are separated. The brown liquor contains lignin, hemicellulose, sodium carbonate, and sodium sulfate. Fifteen percent of the brown liquor is a mixture of solids; the remainder is mostly water, which is evaporated until a solids content of 65-75% is achieved[47][48]. The solids mixture is sent to a recovery boiler where the organic material is combusted. The heat from this reaction is used to generate steam for process energy and to provide sufficient energy to convert the sodium sulfate back into sodium sulfide. A molten mixture of sodium sulfate and sodium carbonate is sent to a causticizing plant where sodium carbonate is converted to NaOH.

The higher heating value is defined as the amount of heat energy released by a specific quantity of material initially at 25° C and bringing the combustion products including water back to 25° C[49]. This heating

and Paper Manufacturing Industry. October 2010. [Cited 28 February 2015] Available from http://www.epa.gov/nsr/ghgdocs/pulpandpaper.pdf

[47] Tran H, Vakkilainnen EK. The craft chemical recovery process. Pulp and Paper Centre University of Toronto. Poyry Forest Industry Oy. [Cited 21 February 2015] Available from http://www.tappi.org/content/events/08kros/manuscripts/1-1.pdf

[48] Northern Pulp Nova Scotia Corporation. Kraft Process at Northern Pulp. [Cited 28 February 2015] Available From http://northernpulp.ca/img/pdf/KraftProcess-NorthernPulp.pdf

[49] Oak Ridge National Laboratory. Lower and Higher Heating Values of Gas, Liquids, and Solid Fuels. Biomass Energy Data Book 2011 [Cited 28 February 2015] Available From

value can only be measured in laboratory settings but is a useful way to compare different fuels. The lower heating value is the amount of heat released by combusting a specific amount of material at 25° C and bringing the combustion products back to 150° C assuming the latent heat of vaporization of water is not recovered. The latent heat of vaporization for water is 2,257 kJ/kg[50]. Black liquor has a higher heating value range depending on its specific constituents of 13.4 to 15.5 MJ/kg total dissolved solids[51]. Gasoline and ethanol have higher heating values of 46.5 and 29.8 MJ/kg respectively[49]. The higher heating value for black liquor is much lower than ethanol and gasoline. In practice, black liquor contains water takes a large amount of energy to vaporize, leading to a lower effective heating value. As mentioned above some of the energy released by the combustion of the black liquor also goes to the formation of Na_2S.

The efficiency of the combustion process in the recovery boiler depends on the amount of water in the black liquor. The moisture content needs to be reduced so that the black liquor contains at least 65% solids, with efficient processes achieving 75% solids. Typical boilers can produce 3.5 kg of steam at 90 bar and 490° C for every kg of black liquor. Higher efficiently systems can produce steam at 110 bar and 510° C. A typical plant produces 1,000 tons of pulp per day, representing 1,500 tons of black liquor. Depending on the thermal efficiency of the boiler, this represents 25 to 35 MW of electricity. The highest efficiency systems can achieve 3.8 kg of steam per kg of black liquor, which equates to 0.6 kWh per kg of black liquor solids. The largest recovery boiler in the world is in Hainan, China, which can process 6,000 tons of black liquor solids per day and produce 734,000 kg steam at 84 bar and 480° C per hour.

According to the World Health Organization (WHO), around 3 billion people cook and heat their homes with open fires that burn

http://cta.ornl.gov/bedb/appendix_a/Lower_and_Higher_Heating_Values_of_Gas_Liq uid_and_Solid_Fuels.pdf

[50] The Engineering ToolBox. Fluids-Latent Heat of Vaporization. [Cited 28 February 2015] Available From http://www.engineeringtoolbox.com/fluids-evaporation-latent-heat-d_147.html

[51] Raguskas, Art. Professor School of Chemistry and Biochemistry Georgia Tech. Renewable Bioproducts Institute. [Cited 28 February 2015] Available From http://ipst.gatech.edu/faculty/ragauskas_art/technical_reviews/Black%20Liqour.pdf

primarily biomass[52]. Most of these people live in poor countries and spend a considerable amount of time procuring fuel for these fires. Around 4.3 million people die each year as a result of diseases caused by indoor air pollution directly related to indoor burning with 50% of pneumonia fatalities in children younger than 5 years old being caused by poor indoor air quality. Of the 4.3 million fatalities from inefficient burning of solid fuels, 12% are due to pneumonia, 34% stroke, 26% heart disease, 22% COPD, and 6% lung cancer.

The main air pollutant from burning wood fuels is particulate matter[53]. The Environmental Protection Agency (EPA) splits particulate matter into two categories: PM_{10} and $PM_{2.5}$. PM_{10} are particles greater than 2.5 μm in diameter but smaller than 10 μm in diameter, and are called coarse inhalable particles[54]. $PM_{2.5}$ are all particles less than 2.5 μm in diameter, and are called fine particles. In the United States, the EPA under the Clean Air Act regulates atmospheric concentration of particulate matter. Fine particulate matter has the worst human health impact because it penetrates deep into the lungs and can even enter the bloodstream. Fine particulate matter is also the cause of haze, which affects the aesthetics of the air and surrounding scenery. As of February 3, 2015 the EPA has set new standards for new residential wood heaters. This includes wood stoves, residential boilers, and pellet stoves. Emissions are limited to 4.5g particulate matter per hour of operation and will drop to 2.5g particulate matter after 5 years. All residential wood heaters have to be certified by the EPA in order to be sold in the United States[53].

With air quality regulations, biomass can be a clean source of energy for residential heat. With inefficient stove and fireplace systems, it can become a hazard to those living in the house and others in the area. The WHO found that having open fires in residential areas increase

---

[52] World Health Organization. Household Air Pollution and Health. [Cited 28 February 2015] Available From http://www.who.int/mediacentre/factsheets/fs292/en/

[53] United States Environmental Protection Agency. Fact Sheet: Summary of Requirements for Woodstoves and Pellet Stoves. [Cited 28 February 2015] Available From http://www2.epa.gov/residential-wood-heaters/fact-sheet-summary-requirements-woodstoves-and-pellet-stoves

[54] United States Environmental Protection Agency. National Ambient Air Quality Standards. [Cited 28 February 2015] Available From http://www.epa.gov/air/criteria.html

the particulate matter to 100 times the recommended level. The EPA annual ambient average standard for $PM_{2.5}$ is 12 $\mu g/m^3$.

Example Problem

Develop a model of smoke concentration as a function of time, C(t), in a house after a brief kitchen fire. The volume of the house, V, is 100 m³, the air flow rate out of the house, Q, is 25 m³ per hour, and the initial particulate concentration, C_1, is equal to the EPA $PM_{2.5}$ concentration of 1,200 $\mu g/m^3$. Let m be the mass of the particulate matter in the air so the accumulation of m in the air over time is give by:

$$Accumulation = in - out + generation - consumption \tag{3.6}$$

$$\frac{dm}{dt} = 0 - QC + 0 - 0 \tag{3.7}$$

Or in terms of mass per unit volume

$$\frac{d\left(\frac{m}{V}\right)}{dt} = \frac{dC}{dt} = -\frac{QC}{V} \tag{3.8}$$

To solve, separate variables and integrate, applying appropriate boundary conditions.

$$\frac{dC}{C} = -\frac{Q}{V}dt = \int_{C_1}^{C(t)} \frac{dC}{C} = -\int_0^t \frac{Q}{V}dt$$

$$C(t) = C_1 e^{\frac{-Qt}{V}} \tag{3.9}$$

Which gives

$$C(t) = \frac{1{,}200\mu g}{m^3} e^{\frac{-0.25}{h}t}$$

Increasing energy production from harvesting woody biomass is not a sustainable solution to decrease the use of fossil fuels[55], according to Schulze, et.al. Direct harvest for biomass results in

---

[55] Schulze, E.-D., Körner, C., Law, B. E., Haberl, H. and Luyssaert, S. (2012), Large-scale bioenergy from additional harvest of forest biomass is neither sustainable nor greenhouse gas neutral. GCB Bioenergy, 4: 611–616. doi: 10.1111/j.1757-1707.2012.01169.x Available From http://onlinelibrary.wiley.com.ezproxy.proxy.library.oregonstate.edu/doi/10.1111/j.1757-1707.2012.01169.x/abstract

younger forests, lower biomass pools, decreased soil quality, and loss of ecosystem function. Studies have shown that 15-25% of global energy production could come directly from biomass by 2050. For instance the United States is mandating increased use of sustainable transportation fuels through 2022 this includes fuels derived from biomass such as ethanol. In 2005 the European Commission adopted the Biomass Action Plan and the Strategy for Biofuels, both try to increase biomass supply and demand. With an increased focus on biomass for energy it is important to understand some of the limitations, such as how biomass energy effects food production, how carbon neutral it is, and sustainability in the long term.

The assumption that all biomass energy is carbon neutral is not true[55]. Biomass stores carbon when it is living and fossil fuels are used in the management, harvesting, and processing of biomass for energy. Recent life cycle analyses done on forests in the Pacific Northwest show that thinning forests to reduce wildfire risk and using the thinned material for energy production actually emits more carbon dioxide than periodic natural fires. Fire acts as a natural disturbance in forest ecosystems and some species actually require fire to reproduce. There are currently policies in place that promote forest thinning for wildfire reduction. Under these policies the benefits of fire and increased carbon dioxide production are not taken into account. There is also policy that promotes thinning in mesic forests, which are wet temperate forests that historically have long fire cycles. It has been shown that because the fire cycles are so long, it is not sustainable to thin for the sole purpose of bioenergy.

The reason that biomass from forests is not sustainable is because forests took centuries to accumulate their carbon biomass. It exists in the living trees and plants, in the soil, and in interactions between different forms of life. By harvesting forest biomass on short-term scales the soil nutrients are changed, the ecosystem is altered, and there is not sufficient time for these systems to recover on a decadal time scale. It has been shown in forests there is a greater potential to store greenhouse gas than there are potential reductions in emissions by replacing fossil fuels with woody biomass. Another interesting phenomena of increased biomass use is a pulse in greenhouse gas emissions. Greenhouse gases are released when the biomass is converted to energy, and it takes decades for that carbon to be recaptured by living plants and animals. For instance, if we started using only biomass for energy we would be emitting more greenhouse gases than growing plants could take up. Even if growing plants were able to accumulate carbon dioxide at the rate we are using it now, we would simultaneously be taking nutrients out of the soil and altering the ecosystem, so that as time progressed, plant growth would become less productive.

In order for biomass energy to become competitive with fossil fuel energy there needs to be technological advances in the manufacture of cellulosic ethanol or incentives such as tax cuts or subsidies. Incentives in the forest products industry can often have unintended consequences. In Germany under the Common Agricultural Policy of the European Union, biomass is subsidized[55] so bioenergy became more common and demand for woody biomass increased. Schulze, et. al. argue that because of this policy, woody biomass prices increased by a factor of five from about 10 euros per cubic meter in 2005 to about 45 euros in 2010. This is about 70% of the price of high quality saw logs. This discourages long-term management for high quality timber that requires good soils and a healthy ecosystem. It encourages short turn around rotations with the extraction of the whole tree including the roots and limbs. It has been shown that removing the roots and limbs increases soil erosion and depletes soil carbon at faster rates than traditional management for saw logs. This type of unintended consequence has been seen before when engineered wood products such as plywood and laminated wooden beams became more prevalent, making high quality timber less valuable.

The most productive forests are not the strongest carbon sinks; generally productivity and ability to store carbon are negatively correlated. Some policies call for acquiring 18-20% of energy from biomass. From Schulte, et. al.[55], humans currently use 7% of the net primary production (NPP) of forests on a global scale; in Europe it is higher at 15%. Net primary production is the synthesis of organic compounds from carbon dioxide. It is estimated that if we were to rely on woody biomass for 20% of our energy production, 18-20% of global NPP would need to be utilized. This represents about 60% of harvestable biomass available globally. Harvestable biomass includes the larger roots, trunks, and limbs. Total NPP includes leaves that are produced and either fall off or are consumed by animals; there are also smaller roots and associations with fungi in the soil that represent biomass that can't be recovered. Sixty percent of harvestable biomass is an overestimation due to location and accessibility of forests and policy that protects forests in conservation areas. For instance, forests high on mountains are hard to harvest, and wilderness areas set aside by countries are inaccessible for harvest.

Overall woody biomass is not a sustainable source to power the entire globe. It is possible to provide up to 20% of our energy needs, but it would have major adverse effects on the environment and our ability to continually produce biomass. Before the decision is made to rely on biomass for large amounts of energy, the big picture needs to be investigated. Life cycle analysis and research should be used to quantify the amount of greenhouse gas emissions both directly and indirectly associated with new energy sources to determine if they truly are renewable. We should continue to use forest products, and practice management strategies that balance long-term sustainability and ecosystem health with productivity. There will always be residuals from the industry that can be used for energy.

Ethanol

Humans have been fermenting grains for thousands of years mostly as a means of preserving food. Ethanol prices have increased due to government subsidies, currently about 36% of corn goes to fuel production[56]. In the United States, 90.6 million acres are devoted to corn, and 14.2 billion bushels were harvested in 2014. To produce ethanol for fuel in 2014, 5.175 billion bushels were used. In 2013, the United States produced 13.3 billion gallons of fuel ethanol and consumed 13.2 billion[57]. For every bushel of corn about 2.8 gallons of ethanol is produced. This means that virtually all fuel ethanol in the United States comes from corn.

Ethanol production is very similar to the beer brewing process; simple sugars in the grain are broken down by yeast into ethanol and carbon dioxide. The dry-grind ethanol production process

[56] United States Department of Agriculture Economic Research Service. Corn Background. [Cited 28 February 2015] Available From http://www.ers.usda.gov/topics/crops/corn/background.aspx
[57] United States Energy Information Administration. How much Corn Ethanol is Produced in the United States? [Cited 9 March 2015] Available From http://www.eia.gov/tools/faqs/faq.cfm?id=90&t=4

produces 70% of the industrial ethanol produced by the United States[58]. In this process the corn is sent through a hammer mill and the particle size is reduced to less than 4 mm. Enzymes are added along with water to make a slurry. The slurry is cooked and agitated by injecting steam and mixing, breaking down the starch into smaller particles. This processes is called liquefaction. The slurry is now called mash, and is cooled to 30° C. Then, another enzyme, glucoamylase, is added. This enzyme breaks down starch molecules into glucose in a process called sarccharification. Yeast is added to the mash, and the simple sugars are broken down into ethanol and carbon dioxide. This is called fermentation, which lasts about 48 hours. Here is the reaction for ethanol production from glucose:

$$C_6H_{12}O_6 \rightarrow 2CH_3CH_2OH + 2CO_2$$

(3. 10)

After fermentation, the slurry is 8-12% ethanol by weight, and it goes to the distillation process where the lower boiling point of ethanol compared to water is utilized to separate the ethanol. The limit in purity of distillation is about 92% ethanol 8% water. Molecular sieves are used to remove water to an ethanol purity of 99%. The solid residuals are separated from the liquid and turned into animal feed. Most of the residual liquid is sent to the beginning of the process to be used in the next batch. The carbon dioxide is usually vented to the atmosphere, but some plants sell it for drink carbonation or for dry ice.

Some Americans have expresses concerned with the amount of corn being used for fuel production. The Federal Renewable Fuel Standard mandates that corn ethanol become a greater fraction of gasoline over the next decade[59]. In 2007, corn prices doubled as a result of ethanol incentives, this increasing food prices for corn, milk, eggs, and meat. Corn is the most important animal feed and represents 94% of all animal feed in the United States. These incentives have increased the amount of land used to grow corn, which means more natural gas is used to make fertilizers, more land needs to be irrigated, and natural habitat is

---

[58]Purdue University Extension Service. Bio Energy. How Fuel Ethanol is Made From Corn. [Cited 28 February 2015] Available From
https://www.extension.purdue.edu/extmedia/id/id-328.pdf
[59] Conca, James. It's Final – Corn Ethanol is of No Use. Forbes. [Cited 28 February 2015] Available From http://www.forbes.com/sites/jamesconca/2014/04/20/its-final-corn-ethanol-is-of-no-use/

converted to cornfields. It is estimated that 25 gallons of ethanol represents enough corn to feed 1 person for 1 year, therefore if the United States' ethanol production changed to food production it could feed 0.5 billion people.

Anaerobic Digestion

Anaerobic digestion is the conversion of organic material to methane and carbon dioxide by anaerobic microorganisms. Anaerobic digestion has been used since the early 1900's in wastewater treatment plants to stabilize sludge. Sludge is the semisolid mixture of sewage left after settling, while stabilization reduces pathogens, minimizes offensive odors, and eliminates the potential for putrefaction. Anaerobic digestion can be a means of breaking down all types of organic matter from food waste, solid landfill waste, and municipal sewage waste. We will explore harvesting energy from anaerobic digestion by focusing on the wastewater treatment process and extraction of methane from landfills.

For municipal sewage, sludge from other processes at the wastewater treatment plant is thickened by removing water that is over 90% of the sludge composition. The thickened sludge is pumped into a digester that is a series of at least two tanks that are kept free of oxygen. The first tank is kept at about 35° C and is mixed. The residence time is typically 10-20 days. The second tank is not mixed and the solids are allowed to settle out of solution. Methane and other gases are collected from the headspace in these tanks. The supernatant is sent to the head of the treatment plant because it is still contaminated. The solids are difficult to dispose of because they typically have elevated levels of toxins such as heavy metals. Currently solids are sent to the landfill, incinerated, or used as a soil amendment if toxin levels are low.

Digester gas is typically 60-65% methane and 30-35% carbon dioxide with small amounts of hydrogen, nitrogen, and hydrogen sulfide. This represents 28 L of gas per person per day for the population the plant services. A simplified view of what happens in the digester is that complex organic molecules are broken down into acetic acid by microorganisms called acetogens. Methanogens produce methane and carbon dioxide from the acetic acid and carbon dioxide through the following reactions:

$$CH_3COOH \rightarrow CH_4 + CO_2 \qquad (3.11)$$

$$CO_2 + 4H_2 \rightarrow CH_4 + 2H_2O$$

$$(3.12)$$

Thirty million tons of food-waste are sent to landfills in the United States, a fact that has recently be the subject of talk show ridicule. This represents 18% of total solid waste[60]. Food waste has three times as much methane production potential as animal and human waste. The EPA and other partners are suggesting that food waste be diverted from landfills and sent to municipal anaerobic digesters. This would be ideal because the infrastructure is already in place and the expertise to maintain digesters' operation is available. Also, the solid residuals are much lower from food waste than from sludge.

The wastewater treatment plant in Eugene, Oregon processes between 34 and 227 million gallons of water per day depending on the weather[61]. To treat this much water, it produces enough sludge that the city needs 3 anaerobic digesters each with a capacity of 3 million gallons. The methane produced goes to heating the anaerobic digesters and to producing electrical power through internal combustion engines operating generators. Eugene has a population of about 160,000 people, and the anaerobic digesters are able to produce enough methane to produce 0.8 MW of electricity. This is only a portion of the treatment plant's electricity demand.

Anaerobic digestion is also a source of methane in landfills. In the United States, 136 million tons of municipal solid waste is deposited in landfills[62] It is estimated that for every one million tons of municipal waste, 432,000 standard cubic feet of methane is produced The composition of landfill gas is 50-55% methane and 45-50% carbon dioxide. Of the total methane emissions in the United States, 18.2% comes from landfills Methane is 25 times more efficient as a greenhouse gas than carbon dioxide.

---

[60] United States Environmental Protection Agency. The Benefits of Anaerobic Digestion of Food Waste at Wastewater Treatment Facilities. [Cited 28 February 2015] Available From http://www.epa.gov/region9/organics/ad/Why-Anaerobic-Digestion.pdf

[61] City of Eugene Oregon. Wastewater Treatment Process. Plant Operations Overview. [Cited 28 February 2015] Available From https://www.eugene-or.gov/index.aspx?NID=454

[62] United States Environmental Protection Agency. Landfill Gas Energy Basics. [Cited 28 February 2015] Available From http://www.epa.gov/lmop/documents/pdfs/pdh_chapter1.pdf

During the first year waste sits in a landfill, aerobic bacteria decompose the organic matter. As more waste is piled on top the aerobes use up the oxygen and anaerobic conditions form, usually after the first year anaerobic digestion process is very similar to the wastewater process discussed, except the time scale is much longer. After the methanogens start producing methane, the steady state methane production lasts for 20 to 30 years

Landfill gas collection systems can be set up for about $25,000 per acre. Collection begins after a cell at the landfill is full and is capped off. The most common type of collection involves drilling vertical wells throughout the landfill and installing perforated pipes In deep landfills horizontal pipes are laid in trenches as waste is added to aid in pipe contact with the waste The gas is pulled from the pipes with a blower and is conditioned by removing moisture and other impurities

Seventy percent of landfill gas energy projects run the gas through internal combustion engines to generate electricity. They are easily maintained and more can be added as capacity is expanded. Internal combustion engines are typical at sites that generate less than 5 MW For landfills that generate more than 5 MW; gas turbines are common; for the biggest operations steam-boiler systems are used

Biomass Conclusion

The human population will probably reach between 10 and 14 billion over the next 50 years. Even with the current world population of just over 7 billion people it is already a challenge to both feed humans and wildlife and maintain an environment worth living in. With increases in human population and standard of life, humans need to take advantage of all types of sustainable energy. It is not possible to rely solely on one source of energy due to limitations in supply of materials, land use, and amount and type of waste.

Biomass has been an important source of energy for humans for thousands of years. At times in our history, biomass, especially wood, has been the most important source of energy. Biomass continues to be important today, though we have to be careful with how we utilize it. Biomass energy is widely available but conventional methods of production are not able to keep up with the amount of energy demand. Land used for biomass production often competes with land used for food production because the most productive lands are ideal for both food and energy production. As carbon dioxide levels continue to increase in the atmosphere, the use of living biomass as a sink of carbon also starts to become a factor to consider. Some sources of biomass such as forests, can be better used as a carbon sink than a source of energy.

Problems

The Brownsville pellet mill produces 84,600 tons of pellets at 2.5% moisture annually from incoming sawdust having a moisture content of 51% by mass. For the previous year it took on average 1,893 Btu to remove 1 pound of water from the material. Lab results from Douglas fir show an available 8,540 Btu per pound of sawdust with 5% moisture.

a) How much wood fuel is used annually in the drying process assume a final moisture content after the drying process is 5%?

b) Using the combustion equation for Douglas fir from the text, how many kg of carbon dioxide are released by 1 kg of Douglas fir?

c) Using the information in example problem 1. How long until the $PM_{2.5}$ drops below the acceptable EPA ambient concentration of 12 $\mu g/m^3$?

d) Complete combustion requires that the products be water and carbon dioxide. What is the chemical formula for complete combustion of ethanol CH_3CH_2OH with oxygen gas? The lower heating value for ethanol is 26.9 MJ/kg [4]. How much heat is released by completely combusting 1 kg of ethanol? How does this value compare to the heating value for Douglas fir used in question 1.

e) A paper mill produces 1,500 tons (1.4 million kg) of black liquor per day, how much power can they produce with the Rankine cycle discussed in example problem 2? Assume that for every kg of black liquor they can produce 3.5 kg steam. Can we ever achieve this amount of power in real life?

Problem Solutions
Problem a

$$\frac{84,600 \text{ tons Pellets}}{\text{year}} \times \frac{0.957 \text{ tons wood}}{\text{tons pellets}}$$
$$= \frac{82,485 \text{ tons dry wood}}{\text{year}}$$

$$\frac{(82{,}485 \text{ tons dry wood})}{\text{year}} \times \frac{1 \text{ ton sawdust}}{0.49 \text{ tons wood}}$$
$$= \frac{168{,}337 \text{ tons sawdust}}{\text{year}}$$

$$\frac{16{,}8337 \text{ tons sawdust}}{\text{year}} \times \frac{0.51 \text{ tons water}}{\text{ton sawdust}}$$
$$= \frac{85{,}852 \text{ tons water}}{\text{year}}$$

$$85{,}852 \text{ tons water} \times 0.05$$
$$= 4{,}293 \text{ tons of water left after drying}$$
$$85{,}852 \text{ tons water} - 4{,}293 \text{ tons water}$$
$$= \frac{81{,}559 \text{ tons of water removed}}{\text{year}}$$

$$\frac{81{,}559 \text{ tons water removed}}{\text{year}} \times \frac{1893 \text{ Btu}}{1 \text{ lb water removed}}$$
$$\times \frac{1 \text{ ton}}{2{,}000 \text{ lb}} = 3.0878 \times 10^{11} \frac{\text{Btu}}{\text{year}}$$

$$3.0878 \times 10^{11} \frac{\text{Btu}}{\text{year}} \times \frac{1 \text{ lb sawdust}}{854 \text{ Btu}} \times \frac{1 \text{ ton}}{2000 \text{ lb}}$$
$$= 18{,}079 \text{ tons sawdust}$$

So 18,079 tons of sawdust at 5% moisture is burned each year to produce 84600 tons of pellets at 2.5% moisture.

Problem b

$$1 \text{ kg Doug Fir} \times \frac{1 \text{ kmol Doug Fir}}{22.75 \text{ kg Doug Fir}} \times \frac{1 \text{kmol CO}_2}{1 \text{ kmol Doug Fir}}$$
$$\times \frac{44 \text{ kg CO}_2}{1 \text{ kmol CO}_2} = 1.93 \text{ kg CO}_2$$

Problem c

$$t = -\frac{V}{Q} \ln \left(\frac{C(t)}{C_0} \right)$$

$$t = \frac{(-100 \ m^3)}{\dfrac{25 \ m^3}{h}} \times \ln \left(\frac{\dfrac{12 \ \mu g}{L}}{\dfrac{1200 \ \mu g}{L}} \right) = 18.4 \ h$$

Problem d

$$CH_3CH_2OH + 3O_2 \rightarrow 2CO_2 + 3H_2O$$

$$1\text{kg}_{\text{ethanol}} \times \frac{26.9 \text{ MJ}}{\text{kg}_{\text{ethanol}}} = 26.9 \text{ MJ}$$

$$26.9 \text{ kg}_{\text{ethanol}} \times \frac{947 \text{ Btu}}{\text{MJ}} \times \frac{0.453 \text{ kg}}{1 \text{ lb}} = 11,540 \frac{\text{BTU}}{\text{lb}}$$

$$\frac{11540 \left(\frac{\text{Btu}}{\text{lb ethanol}}\right)}{8540 \left(\frac{\text{Btu}}{\text{lb wood}}\right)} = 1.35$$

Ethanol has a heating value that is 1.35 times greater than Doulas Fir at 5% moisture.

Problem e

$$1.4 \times 10^6 \text{ kg black liquor} \times \frac{3.5 \text{ kg steam}}{1 \text{ kg black liquor}}$$
$$= 4.9 \times 10^6 \text{ kg steam}$$

$$\frac{580 \text{ kJ}}{\text{kg steam}} \times \frac{4.9 \times 10^6 \text{ kg steam}}{\text{day}} \times \frac{1 \text{ MJ}}{1000 \text{ kJ}} \times \frac{1 \text{ day}}{86400 \text{ sec}}$$
$$= 32.9 \text{ MW}$$

We could never achieve this because we assumed a reversible system.

Chapter 4. Nuclear Reactors
By Pavel A. Grechanuk

Nuclear reactors are found throughout the world and are used in a variety of applications, such as powering satellites, generating electricity, creating medical radioisotopes, and providing process heat for industrial applications. The most common applications of nuclear energy is to generate electricity and to power submarines. In the United States, nuclear power provides for 19% of our electric energy needs and powers over 140 naval vessels with more than 12,000 reactor years of operation on the sea. Nuclear energy is a safe and reliable energy source that produces no greenhouse gas emissions.

Nuclear power plants are almost identical to any other steam driven power plant in the world. A heat source boils water, and the high-pressure steam generated is used to spin a turbine–generating electricity. The hot steam is condensed and the excess heat is transferred to the environment. The principle difference between non-nuclear and nuclear plants is the heat source. In a combustion plant; coal, oil, or natural gas is burned in furnaces to produce steam, whereas in a nuclear plant, the heat comes from the self-sustaining fission chain reaction of uranium.

Nuclear reactors have the capability to produce enormous amounts of thermal energy without releasing carbon dioxide or other pollutants. When a neutron enters the nucleus of a heavy atom, nuclear fission may occur. The large nucleus may become unstable and break apart into two or more lighter elements, called fission fragments, which releases neutrons, gamma radiation, and large amounts of energy. The total mass of the fission products is less than the original atom. This is because a small portion of matter converts directly to energy by the following formulas, where E is energy and c is the speed of light.

$$A + B \longrightarrow C + D$$

$$\Delta M = Mass_A + Mass_B - Mass_C - Mass_D$$

$$E = \Delta Mc^2$$

The mass lost is usually very small, but the energy equivalent represents an enormous amount of heat transfer to the surroundings. Any neutrons that are released may go on to cause subsequent fission reactions, which release even more neutrons. This is a nuclear chain reaction and it is the primary process by which nuclear power plants generate heat.

For a self-sustaining chain reaction to occur, the amount of neutrons being produced must equal the amount consumed. When the number of neutrons is in equilibrium, the reactor is considered critical. Every atom has a neutron capture cross section, which is a probabilistic measure of the cross-sectional area on the atom presented to an incoming neutron. The absorption cross section is highly dependent on the energy of the incoming neutron. The average energy of a neutron resulting from fission is 2 MeV classifying it as a fast neutron, and neutrons with approximately 0.025eV are called thermal neutrons. Neutrons with less energy have a higher probability of being absorbed into a nucleus and causing fission, while high energy fast neutrons have a lower probability of absorption. Fast reactors and thermal reactors are named for the energies of the neutrons that they use. As previously stated, neutrons resulting from fission have an average energy of 2 MeV, thus to maintain criticality in a thermal reactor the neutrons have to be slowed down by a material called the moderator. The moderator in most reactors is either graphite or the liquid coolant.

Fast reactors use unmoderated neutrons, and as a result require a relatively high density of fissionable material to maintain a chain reaction. However, fissions that are caused by fast neutrons release more neutrons than fissions caused by thermal neutrons, meaning that a properly set up fast reactor can breed more fuel than it consumes. Breeder reactors extract nearly all of the energy in thorium and uranium, decreasing the fuel to energy requirements by a factor of 100. Traditional thermal water reactors only extract 1% of the total energy contained in the uranium. After the 1% of the fuel has been burned, the remaining is considered waste and moved to storage. Due to their ability to extract virtually all of the energy out of nuclear fuel, breeder reactors have the capability to satisfy the energy needs of the future. In the oceans on Earth, there is an estimated 9×10^9 pounds of uranium dissolved in the water. It has been proposed that utilizing breeder reactors and seawater uranium extraction facilities together would effectively make nuclear power an unlimited energy source. Research and development into breeder reactors has been stagnant since the 80's due to the extremely low price of uranium fuel.

Every nuclear reactor has to be cooled by a working fluid that carries the heat away from the core. The thermal properties of the coolant vastly affect the design of the reactor. The most commonly used coolant is water, due to the fact that it is very cheap and engineers have been using water as a working fluid in conventional fossil fuel plants for years. The main disadvantage of using water in a reactor is that it must not be allowed to boil. Steam is less dense than water, and does not moderate neutrons as effectively. Therefore, if steam is present in a water-cooled reactor, the reactivity decreases, because the number of moderated neutrons that cause fission decreases.

In a reactor system, the thermal efficiency is a function of the temperature difference between the steam produced and the cooling water. As the temperature difference increases, the thermal efficiency also increases. The critical temperature for water, 375 °C, is the temperature above which liquid water cannot exist. Thus, in order to maintain criticality, temperatures in water-cooled and moderated reactors must not exceed the critical temperature of water. Consequently, coolant temperatures are mostly not allowed to exceed 340 °C, and this temperature limit sets the maximum thermal efficiency of water-cooled reactors at 34%.

The fission rate in a reactor is controlled using control rods that are made out of a neutron poison (boron, silver, cadmium, or indium). These elements are called neutron poisons because they have the ability to absorb a high number of neutrons without undergoing fission. The control rods are inserted into the core to reduce neutron flux, which is the number of neutrons traveling through an area per unit time. The more neutrons you have per unit area, the more fission you have occurring in that area. The neutron flux affects the rate of fission, which affects the amount of heat and steam produced, and therefore the amount of electricity generated. This is the method to control the amount of electricity produced by a nuclear power plant.

Nuclear reactors are different than combustion furnaces in the fact that the reactor keeps producing heat even when shut off. Inserting the control rods into the reactor completely stops the fission chain reaction, but the radioactive decay of the lighter fission fragments continues to generate heat. If you immediately shut down a reactor from full power, the decay heat will be approximately 6.5% of the initial reactor heat. The decay heat generation depends upon the fraction of the numerous fission fragments and their respective half-lives. It is for this reason that heat needs to be constantly carried away from the core, or else a core meltdown is possible.

Brief History

To understand the significance of nuclear technology, one must first understand its origins. In the 19th century, breakthroughs in the rapidly growing field of chemistry led to discoveries that supported the atomic theory of matter. In 1909, Hans Geiger and Ernest Marsden, with the aid of Ernest Rutherford discovered the existence of the nucleus through their gold foil experiment. In 1932, James Chadwick discovered the neutron, and not two years later Enrico Fermi was the first person to bombard a uranium nucleus with neutrons and cause fission. His experiment proved that atoms could be split using neutrons leading to a skyrocket of research into nuclear physics. In 1942, a team of scientists built the first man made nuclear reactor at the University of Chicago. Led by Fermi, the team took less than a year to work out their theories, complete a design, and build Chicago Pile-1.

The first reactor was only the beginning in the development of nuclear technology. During World War II, The United States, fearing that the Germans were developing a nuclear weapon, launched the Manhattan Project. The scale of the project was massive: at its peak employing over 130,000 people and costing an approximate $26 billion in 2015 dollars. The project lead to the construction of world's first atomic bomb, which was used against the Japanese at Hiroshima and Nagasaki to finally end World War II.

The Manhattan Project laid the foundation for the technological applications of nuclear technology, in the form of multiple national laboratories including: Oak Ridge National Laboratory, Los Alamos National Laboratory, Argonne National Laboratory, and many more. These laboratories would be the forefront of nuclear research in the years to come. In 1953, U.S. President Dwight D. Eisenhower gave his famous speech 'Atoms for Peace' to the United Nations, urging the necessity of developing peaceful applications for nuclear technology. The 1954 Amendments to the Atomic Energy Act was passed shortly thereafter, permitting the declassification of top secret reactor technology and motivating expansion in the private sector.

Worldwide the total capacity of nuclear power rose from 1 gigawatt in 1960, to 100 gigawatts in late 1970. In 1973, a world oil crisis occurred, which put pressure on countries to find alternative electrical generation methods. At the time Japan and France heavily relied on oil to generate electricity, 73% and 39% respectively, and it motivated them to invest heavily in Nuclear. Today France generates

70% of its electricity from nuclear, and enjoys some of the cheapest electricity in Europe. Currently, Japan has the capacity to generate over 30% percent of its electrical supply using nuclear. However, following the tsunami that that hit the Fukushima Daiichi plant in March of 2011, they have shut down many of their reactors for safety inspections.

There have been a variety of different and unique designs for power reactors and a multitude of prototypes have been built. The first generation of power reactors, Gen I, used graphite or heavy water as the moderator and natural uranium as the fission fuel. However, today most nuclear reactors use 3-5% by weight ^{235}U, allowing light water and other moderators to be used. Most of the nuclear power that is generated in the world comes from this second generation of nuclear reactors (Gen II) built throughout the 1970s and 1980s. These reactors were designed with a 40-year life expectancy, and are now extending their operating licenses in order to continue producing power.

Gen II

The most common nuclear power plant in the world today uses pressurized water reactors (PWRs) to generate electricity. A PWR operates in the thermal neutron spectrum using water as a coolant and neutron moderator. A PWR utilizes a two-loop water system; in the main loop the water is held at high pressure to keep from boiling as it flows past the core to absorb heat. The hot pressurized water flows to a steam generator where it passes its thermal energy to a non-radioactive secondary loop of water; generating high-pressure steam to spin a turbine and generate electricity. The use of a two-loop system is a standard practice to ensure that radioactivity does not leave the primary loop.

The second most common reactor type is a boiling water reactor (BWR). The difference between a BWR and a PWR is that the BWR only has one loop of water and allows boiling of water in the core. The cooling water boils into steam as it flows through the core to the steam water separator. The dry steam then flows directly to a turbine, is condensed, and pumped back to the reactor. By having a single loop, the overall complexity of the system is decreased, and there is no need for steam generators and other complex equipment that is necessary in a PWR.

GEN III

Currently, the next generation of nuclear reactors (Gen III) is under construction throughout the world. This next generation of reactors incorporates the technological advancements that have been made in nuclear science throughout the past decades. These improvements include passive safety, improved fuel design, increased thermal efficiency, longer operation lifetimes, and standardized design for decreased capital and maintenance costs.

A reactor that is being built right now (2015) is the AP1000, designed and sold by Westinghouse. It is a two loop pressurized water reactor (PWR) that has been highly simplified to increase safety, construction speed, and capital return. It has a power output of 3,400 megawatts thermal with an optimum electrical output of 1,100 megawatts electric. Simplification was a major objective, and as a result the safety systems are significantly less complex than those of a standard Gen II reactor.

The AP1000 has 50% less valves, 35% less pumps, 80% less piping, and is 45% smaller by volume than its predecessors. The fact that this reactor has a much simpler design makes it easier to build, operate, and maintain. The safety of the AP1000 has been significantly improved over its predecessors. As mentioned before, due to the decay heat of the fission fragments, the coolant must flow even when the reactor is shutdown. In the AP1000 the passive cooling system cools the reactor, even when connection to the grid is lost. The system does not need pumps or connection to electricity, but relies on natural convection to cool the reactor. Natural convection describes a process of heat transport, where the fluid motion relies on the density changes in the fluid caused by temperature gradients. As long as the gravity drained water tank on the roof is filled, the reactor will cool itself indefinitely, without relying on pumps or complicated systems. This is a massive safety improvement over the Gen II reactors, which require pumps to keep the reactor cool.

The AP 1000 is the forefront of current buildable reactors, but there is a new generation of reactors currently being researched. They are called the Generation IV reactors, and are not expected to be viable for commercial generation until 2030. When compared to Gen III reactors, the Gen IV reactors boast some impressive benefits: up to 300 times more energy from the same amount of fuel, much broader diversity of fuel options, shorter lived radionuclides, and the ability to use existing waste as a fuel source. These reactors avoid using water as a coolant, and as a result eliminate the risk of steam explosions, hydrogen generation (can lead to a hydrogen explosion), and water contamination. This technology has been in development for years, and research is continuing at universities and national laboratories.

Gen IV

One of the most versatile reactors that is currently being researched is the Liquid Fluoride Thorium Reactor (LFTR) which is a Gen IV molten salt reactor, MSR. MSR's contain the nuclear fuel in the form of a molten salt mixture. Molten salt reactors can be either burner or breeder reactors, in the fast or thermal spectrum. The reactors use either a fluoride or a chloride salt fuel and a variety of fissile and fertile additives. LFTRs exclusively use a thorium fluoride salt to breed thorium into uranium which can undergo fission to produce energy.

In a LFTR reactor the molten salt is pumped to a core, where the fuel achieves critical density and a chain reaction occurs. The molten salt then flows to a heat exchanger, where the heat is transferred to a secondary loop where steam is generated. Due to the design of the LFTR and the molten salt fuel, this reactor system has some very unique advantages.

The first advantage that should be mentioned is operation at low pressure. Standard light water reactors (LWR) use pressurized water as a coolant and neutron moderator. If the high-pressure water escapes through a leak, it instantly flashes to steam. This steam expands to a thousand times the initial volume of the water, meaning the containment building for a light water reactor needs to be a thousand times larger than the reactor vessel to prevent an explosion.

By using a molten salt coolant instead of water, the containment structure only needs to be a little larger than the reactor vessel. This is possible because the molten salt mixture remains liquid at extremely high temperatures. The core of a LFTR is designed to operate at 0.6 MPa, which is comparable to the pressure of the public water system. Additionally, the molten fuel has a very high melting point, thus even a several hundred-degree temperature increase does not cause a significant pressure increase. Even in the case of a core leak, there is not going to be a significant volume change, therefore an explosion within the containment building is not likely. These factors give LFTRs an advantage over light water reactors in terms of inherent safety, smaller size, lower materials use, and lower construction costs.

The next advantage is that LFTRs produce up to 1000 times less transuranic waste than conventional light water reactors. In a typical LWR the ^{238}U in the fuel is transmuted to ^{239}Pu, which has a half-life of 24,000 years. Therefore most of the fuel is only one step away from becoming a long-lived transuranic isotope. In comparison, LFTRs operate via the thorium cycle, which transmutes thorium into ^{233}U. As thorium is a lighter element it requires more neutron captures to transmute to a fissile element. In the LFTR, ^{233}U has two chances to fission, the first as ^{233}U of which 90% will fission, the remainder then has another opportunity as ^{235}U of which 80% will fission. This only leaves 2% of the fuel to go on to become transuranic, which equates to 15 kg per GW-year. This mass is 20 times smaller than the mass produced in a traditional LWR. More importantly the transuranic waste products can be sent right back into the core to fission, which produces more energy and consumes the waste. When the benefits of lower transuranic waste and breeding are utilized together, the production of transuranics is reduced by over a factor of a 1,000 compared to standard LWRs.

Liquid fluoride thorium reactors still produce radioactive elements, but they have a significantly shorter half-life than the ones produced in LWRs. The radiotoxicity of the fission products is mainly due to ^{137}Cs and ^{90}Sr. Cesium has a half-life of 30.17 years, meaning after 30 years the radioactivity is reduced by half. Ten half-lives (300 years) will reduce the radioactivity by a factor of 1,024, making it less radioactive than uranium ore. This is 230,700 years shorter than the time required for LWR waste to reach the same level of radioactivity. More importantly you can separate out the fission products during operation of the reactor, which enables them to be partitioned and stored based on half-life. Within 10 years, 83% of the fission products from a LFTR are stable and can be used in industry or in medical applications.

The final advantage is that LFTRs are very efficient, when compared to their LWR counterparts. Standard LWRs consume less than a percent of the uranium in the fuel, leaving the remaining 99% as waste. With proper reprocessing, a LFTR could consume 99% of the thorium fuel. Due to the high operating temperatures of the reactor, the closed gas Brayton cycle can be used, which boosts the thermal to electric efficiency to 54%. The closed gas Brayton cycle spins a turbine using hot gas as a working fluid instead of steam. The increased efficiency results in a 20% reduction of fuel consumption and fission products production. The two factors of thermodynamic and reactor efficiency combined, mean that 1 ton of thorium in a LFTR produces as much energy as 35 tons of enriched uranium (requiring 250 tons of natural uranium) in a standard LWR, or 4.16 million tons of coal in a coal power plant.

Fusion Power

Almost all life on Earth in one way or another depends on nuclear energy to survive. All of the stars in the universe, along with our sun, are massive nuclear reactors. Stars operate via a fundamentally different nuclear process: fusion. Instead of splitting a nucleus, fusion actually combines two lighter nuclei into one larger nucleus. In order for fusion to occur the kinetic energy of the lighter nuclei must overcome the columbic barrier. The columbic barrier results from the repulsive force felt by like charges (electromagnetic force). Hydrogen is the ideal fuel source for fusion, as it has the smallest charge, and is easiest to fuse.

In a fusion reactor there are two ways to overcome the coulomb barrier: high heat and high velocity. When an atom is heated to the point that it has more energy than its ionization energy, it loses its valence electrons and becomes an ion. When the majority of atoms are ionized, the resulting sea of electrons and ions can be manipulated and confined using strong magnetic fields. This is the process by which scientists hope to achieve and sustain a fusion reaction.

The total power output that can be achieved from a nuclear fusion reaction is governed by the following equation:

$$E_{Recoverable} = \eta(E_{Fusion} - E_{Radiation} - E_{Conduction}) \qquad (4.1)$$

The energy from fusion is not completely recoverable, as the plasma loses energy to the surroundings. Conduction occurs when the ions and electrons impact the walls of the reactor, transferring their kinetic energy to the atoms in the wall. Radiation losses occur when energy leaves the plasma in the form of light. The variable η is the efficiency of the plant at capturing the heat. To achieve more energy out of a fusion reactor than you put in, the fusion reactions have to occur at a high enough frequency to make up for energy losses. The three criteria that must be met are sufficient plasma density, temperature, and containment time. Most research focuses on increasing these three criteria.

Achieving fusion would be a massive technological breakthrough for humanity. The fusion of hydrogen releases the greatest amount of energy per mass, than any fuel source currently in use. The primary fuel deuterium, which exists in the oceans on Earth, is an isotope of hydrogen with two neutrons. The relative abundance of deuterium is low, about 0.015% of all water, but because fusion reactions release so much energy, fusion reactors could supply all of the world's energy needs for millions of years. The excess heat from the reaction could be used to operate a desalination plant, which would provide both drinkable water and more deuterium fuel. Fusion produces no radioactive waste, has minimal nuclear proliferation risks, and could be used to safely power the developing world. Fusion reactors could one-day power space ships that are travelling on interstellar missions outside of our solar system. Fusion energy has the capability of opening a new chapter for humankind, which is why billions of dollars have been spent researching this technology. The first fusion reaction with a positive energy gain is still expected to be decades away.

Space Reactors

Many people don't consider the fact that there are nuclear reactors orbiting Earth. While solar power is the most common form of generating electricity in space, it has its limitations. The intensity of solar light fades off in an inverse square relationship, $I = \frac{1}{D^2}$, meaning if you travel twice as far away from the sun, the intensity drops by a factor of four. The result of this is that for deep space missions, the further your destination is from the sun, the bigger your solar panels need to be to generate the same amount of energy.

Nuclear reactors are especially useful due to their low weight to energy density ratio. Solar panels take up much more space than nuclear systems with the same capacity. Spacecraft that are small and compact are easier to maneuver and orient when precision is needed. Nuclear power could be effectively used for propulsion along with powering life support, making it the ideal energy source for any mission that hopes to leave the solar system.

The most common nuclear reactor in space is called a radioisotope thermoelectric generator or RTG for short. RTGs use a system of thermocouples to convert the decay heat from certain long-lived radioisotopes into electricity by the Seebeck effect. When two metals are joined and a temperature gradient is applied across them, the charge carriers in the metals diffuse from the hot side to the cold side. The movement of charge is the definition of electrical current.

The simple design of the RTG has unique advantages when compared to other electrical generation methods for use in space. First, it has no moving parts, so there is no mechanical wear. Second, RTGs can produce a fairly steady supply of energy for decades. This makes RTGs the most desirable electrical generation source for isolated situations that need a few hundred watts for an extended period of time. RTG's can be found on space probes, satellites, and the Russians have even used them in remote research facilities in the arctic.

Economics

For any real world application of technology, one of the most important aspects is economic feasibility. When talking about the economics of nuclear power, one has to be concerned with three things: principal capital costs, operating costs, and social costs. Nuclear power is cost competitive with other forms of energy generation only when there is no access to fossil fuels. Nuclear electric plants are extremely costly to construct, but inexpensive to operate. Calculating the costs of generating electricity using different technologies is often very challenging, and more often than not it depends on where the plant is being built. Fossil fuels are, and will continue to be, the most economically attractive method of generating electricity, for as long as carbon emissions are tax free; especially in countries where fossil fuels are abundant, like the United States, China, and the Middle East. But, if you address the health, environmental, and social costs of fossil fuels, nuclear power becomes much more attractive.

The capital costs of a nuclear plant are defined as the construction, manufacture, financing, and commissioning costs. The initial construction of a reactor takes the effort of thousands of people using massive amounts of concrete and steel. Many engineers and professionals are employed to set up the systems that provide reactor controls, information communication, ventilation, and cooling. The most common form of measuring costs is dollars per kilowatt-hour. Compared to natural gas or coal, the cost per kWh of nuclear can be two to three times higher. If you take into account the financing costs of building a nuclear plant, the capital costs are dependent on the duration of time it takes to construct the reactor, the mode of financing, and the interest rate.

Policy makers in Washington D.C. have recognized the value of nuclear as an energy source, and have supported the construction of new nuclear plants. The Energy Policy Act of 2005 incentivized the construction of new nuclear plants by providing tax breaks per unit of energy connected to the grid. The act also created a loan guarantee program within the Department of Energy to support technologies that reduce carbon emissions. Certain states have created financing tools, which aid in reducing the capital costs of nuclear plants. An example is the 'construction work in progress', which allows the recovering of interest costs during construction. If the finance costs are not recovered during construction, they compile, and are added to the total capital cost; only to be recovered when the plant begins producing electricity. This practice lowers the overall capital cost of the power plant, in effect reducing the cost that consumers pay for electricity.

The operating costs of nuclear include: the cost of fuel, maintenance, disposing of used fuel, and eventual decommissioning. The cost of natural uranium is extremely cheap, but it has to be enriched and made into fuel pellets before it can be used in a reactor. The U.S. Energy Institute calculated that the fuel cost of nuclear is 14% of the total operating cost, compared to a natural gas plant where 89% of the cost is due to fuel. The relative cheapness and abundance of uranium is one of the factors that dampen interest in fast reactors, which burn up a larger portion of the fuel. In a light water reactor, after the uranium fuel has spent about three years in the reactor, it is moved to a temporary water containment tank. The spent fuel resides there until the radiation levels are low enough to either move it to dry storage or a reprocessing facility. The costs resulting from this process are known as the back end fuel costs. Nuclear energy is unique in the fact, which it has to be concerned with the whole life cycle of the fuel, from mining all the way to waste storage. The costs of spent fuel storage and disposal, account for about 10% of the total cost per kWh of nuclear. The final decommissioning costs of nuclear reactors account for no more than 5% of the cost per kWh.

Societal Aspects

Societal costs are not included in the construction or operation of any power plant, but are still paid by the general community. Societal costs are defined as costs resulting from damage to the environment and human health. ExternE was a major European study conducted in 2001 that focused on the societal costs of various energy generation methods. It concluded that in monetary values, the societal price of coal is ten times greater than that of nuclear energy. If societal damage were incorporated into the price of electricity, nuclear would be three times as cheap as coal. The only way for renewables and nuclear to compete on a fair playing field with fossil fuels is for the federal government to instate a carbon tax. Currently Colorado, California, and Maryland are the only states that have passed carbon tax laws.

In 2013 the U.S. National Aeronautics and Space Administration (NASA) released a qualitative report named, 'Coal and Gas are Far More Harmful than Nuclear Power', comparing the effects of nuclear, coal, and gas on the climate and human mortality. The paper found that without nuclear power, it would be extremely difficult to combat global warming. This is due to the fact that if nuclear power did not exist, the energy produced by it would have come from fossil fuels instead. Using historical data and mortality factors from scientific papers, they found that despite the three major nuclear accidents that have happened, nuclear power has prevented over 1.8 million deaths worldwide between 1971 and 2009. This amounts to thousands of lives saved for each death caused by nuclear.

NASA also calculated that nuclear energy has prevented a release of 64 gigatonnes of greenhouse gases in the same time period. This is equivalent to the past thirty-five years of emissions from coal fired plants in the U.S. They calculated that natural gas, being cleaner than coal, causes 40 times more deaths per unit energy than nuclear. The conclusion of their report was that: nuclear energy, despite having unique challenges, needs to be significantly expanded in order to minimize the impacts of unabated global climate change and air pollution caused by fossil fuels.

Nuclear power is an amazing technological achievement, and it has the capability to change the world. Why does energy matter? Because, the way society decides to generate energy has vast societal, ecological, and political implications. Nuclear power is a way to generate clean and safe electricity for the world. On Earth there are still

1.3 billion people who live without electricity, and these people have yet to experience the societal benefits of electricity. The energy portfolio that will produce electricity for the remaining 20% of the world will drastically affect the future of the Earth. It is my hope that in the following years the global community will strive towards a more environmentally friendly energy portfolio. Locally in the U.S., the best way to expand nuclear energy and renewables is to impose a carbon tax, which would make the energy market a more equal playing field. Making the fossil fuel industry responsible for their waste stream, would allow nuclear and renewables to be cost competitive and incrementally displace fossil fuels.

Chapter 5. Solar
By Hannah Bulovsky

Solar energy is energy from the sun--energy is produced nuclear fusion. The fusion of compressed hydrogen atoms releases a very large amount of energy, which radiates out toward Earth. This energy is used by plants for photosynthesis, and is necessary for life. Electricity production requires a heat source and a cold sink as a driving force. For solar energy, the heat source is the sun and the cold sink is the solar panel, heat transfer fluid, or buildings absorbing sunlight for electricity production or heat. Renewable resources are defined as resources that are naturally replenished. Sun, wind, and hydroelectric power all fall into this category. Solar power is a carbon-free energy source, meaning no carbon dioxide (CO_2) is released during energy production.

There are two main types of solar energy: active and passive. In active processes, solar radiation can produce electricity using photovoltaic technology, by concentrating heat for thermal use, or by producing steam to rotate a turbine. This turbine generates electricity, which can provide power to nearby buildings. Passive solar provides light and heat to buildings based on their design.

Solar photovoltaic (PV) is a form of active solar, and the most widely used form of solar power. PV cells use semiconductors, or materials that can be used to control the flow of current, to collect electricity. A single cell can produce around 2 watts of power, which is not enough power to run an electric watch or calculator. Many PV cells can be arranged into a pattern called an array, and these arrays make up solar panels. As sunlight hits a cell, an electric charge is created within the semiconductor. This electric charge can be transferred through a circuit and power electric devices. Many of these cells make up the panels that collect electricity on homes and solar farms.

Another active solar process is using solar energy for heating. Heat can be concentrated in water pipes to produce hot water for use or for heating buildings. Thermal electric is similar to solar heating. Heat from the sun is concentrated and used to heat water, but in thermal electric applications the water is turned to steam. Enough steam is produced to power a turbine, which provides electricity (EPA).

The orientation and window layout of a building-using passive solar are designed so their electric heat and lights are supplemented by solar energy. Houses that use passive solar for temperature control may have larger south facing windows and smaller windows throughout the

71

rest of the house. This design allows heat from the sun to warm the house from the south, while preventing heat losses through other windows. In the warmer months, the large windows can be shaded to keep the house cool. These homes are usually very well insulated, to keep heat in in the winter and out during the summer. Homes built using passive solar and with energy efficiency in mind can use up to 60% less energy than typical homes without sacrificing comfort (*Before the Lights Go Out*).

The History of Solar

Passive solar has been used to heat homes for centuries. Between the 1st and 4th century A. D., Roman bathhouses employed passive solar by having south facing windows to take advantage of the heat from sunlight. Native Americans also built southward facing houses to utilize passive solar for heating around 1200 A. D. The first office building to use solar water heating and passive solar was designed in the mid-1950's by Frank Bridgers.

In the 7th century B. C. solar thermal energy has been used since to light fires. Mirrors and glass were used to concentrate sunlight onto tinder to start fires for light and cooking. Solar heating became more widely used in the second half of the 18th century. In 1767, Horace de Saussure, a Swiss scientist, created an insulated 'hot box' that absorbed solar energy to cook food, and in 1816 Robert Stirling patented an engine that concentrated thermal energy to produce electricity. Stirling's engines are still used today in thermal energy plants. In 1986, the world's largest solar thermal facility was built in Kramer Junction, California. Mirrors focus the sun's energy onto a pipe containing a heat transfer fluid. The heated fluid then transfers the energy into water to produce steam, which generates electricity by powering a turbine.

Edmond Becquerel discovered the PV effect in 1839. In 1876 William Grylls Adams and Richard Evans Day found that selenium produced electricity when exposed to light. Satellites with arrays of PV cells were launched in 1958 after the development of silicon PV cells that could better resist radiation. Although commercial solar power was not successful at this time, it was used to effectively provide power to satellites. Solar cells began powering lighthouses, warning lights on offshore gas and oilrigs, and railroad crossings during the 1970s. The first PV cell powered residence was built in 1973.

In 1977, the U. S. Department of Energy founded the Solar Energy Research Institute, which later became the National Renewable Energy Laboratory (NREL). That same year, the amount of power produced by PV cells surpassed 500 kilowatts. One of the main barriers to widespread use of solar power is efficiency, and considerable research efforts have gone into minimizing the energy lost during the conversion of sunlight to electricity.

Measures of Solar Cell Performance

Solar cell efficiency, η, is the ratio of amount of energy produced by the cell to the energy that reaches the surface of the cell. The maximum power output of the cell, P_m, is divided by the light input, E, and the cell area, A, shown in Eq. 5.1.

$$\eta = \frac{P_m}{EA} \times 100\% \qquad (5.1)$$

P_m in watts (W) is the power output of the cell at maximum production, which occurs when sunlight is shining straight onto the cell, or when the angle of incidence is zero. The amount of power a cell can produce from a given amount of sunlight depends on the angle the light strikes the cell. The most light can be absorbed when the light hits the cell straight on, which means this equation for efficiency gives the maximum possible efficiency of the cell. The light input, E, is the amount of sunlight that hits the cell per area of cell. E is usually measured in units of watts per square meter. A is the area of the surface of the cell, measured in square meters. A typical value for E when calculating solar cell efficiencies is 1,000 W/m^2.

The maximum power output of the cell is calculated using the short circuit current, I_{sc}, and the open-circuit voltage, V_{oc}, shown in Eq. 5.2.

$$P_m = I_{sc}V_{oc} \qquad (5.2)$$

The open-circuit voltage, V_{oc}, measured in volts (V) is the voltage across a terminal that builds when there is no current flowing, which occurs when the circuit is open. The short circuit current, I_{sc}, measured in amps (A), is the current that flows when there is zero resistance. This means that the terminals are connected. The short circuit current density, J_{sc}, can be calculated in using the short circuit current density and the solar cell surface area, A, shown in Eq. 5. 3.

$$J_{sc} = \frac{I_{sc}}{A}$$

(5. 3)

The short circuit current density can be used to compare the performance of solar cells.

Example 1: What cell area (in m^2) is needed for a 20% efficient solar cell with an open-circuit voltage of 0.5 V, and a short circuit current of 4 A? Use a typical value for light input.

Solution:

Use Eq. 5.2 to calculate P_m:

$$P_m = I_{sc}V_{oc}$$

$$P_m = (4\ A)(0.5\ V)$$

$$P_m = 2\ W$$

Rearrange Eq. 5.1 to determine the area:

$$A = \frac{P_m}{E\eta}$$

$$A = \frac{2\ W}{(1000\ \frac{W}{m^2})(0.20)}$$

$$A = 0.01\ m^2$$

A 0.01-m2 cell is needed to produce 2 W of energy. For a round solar cell, this gives a cell diameter of 0.11 m, or about 4.5 inches.

In 1985, the first silicon solar cells with greater than 20% efficiency were developed at the University of South Wales. In 1999, the NREL and Spectrolab, Inc developed a three layer PV cell that converted 32.3% of the sunlight into electricity. In 1992, the University of Florida developed a thin-film cadmium telluride PV cell that was over 15% efficient. By 1999, the NREL developed a thin-film cell that was

18.8% efficient. In 1994, gallium indium phosphide and gallium arsenide cells were developed by the NREL that exceeded 30% efficiency in laboratory conditions.

How Photovoltaic Solar Cells Work

There are several materials that can be used to collect solar energy. Charles Fritts described the first solar cells made from selenium wafers in 1883, and in 1918 Jan Czochralski developed a method of growing single-crystal silicon, which can be used in PV cells. Cadmium sulfide (CdS) was found to exhibit photoelectric effects in 1932. In 1954, Bell Labs developed the first silicon PV solar cell capable of producing enough electricity to power electrical equipment. Cadmium telluride, copper indium gallium selenide, gallium arsenide, and amorphous (non-crystalline) silicon have also been used to make solar cells.

Materials used for solar cells must be semiconductors. As mentioned above, a semiconductor is a material that can conduct electricity. In solar PV cells, when sunlight strikes the semiconducting wafers, electrons are dislodged and move to fill vacancies left by other electrons. If the wafers are wired together, these electrons will travel, creating an electric current.

Many solar cells use a combination of materials to achieve the highest possible flowing current. When the electrons are dislodged, they leave behind a hole that can be filled by another electron. Some materials tend to develop more holes that are unfilled than others. These materials are called p-type semiconductors. Materials that develop an excess of free electrons are called n-type semiconductors. PV cells use both n-type and p-type semiconductors, separated by a material known as a p/n junction. The n- and p-type materials create an electrode, allowing the flow of electrons from one side to the other. This creates the current necessary for electricity production (Solar Cell Central).

Band gap energy is the energy required to dislodge an electron, freeing it to flow and produce electricity. The light hitting the cell must be equal to or greater than the band gap energy to free the electrons. The band gap energy of crystalline silicon is 1.1 electron volts (eV), where 1 eV is equivalent to 1.602×10^{-19} joules. One eV is the amount of energy gained by one negatively charged electron moved across an electric potential of one volt.

Example 2: How many joules are in 6 eV?

Solution:

$$6 \text{ eV} \left(\frac{1.602 \text{ x } 10^{-19}\text{J}}{1 \text{ eV}} \right) = 9.6 \text{ x } 10^{-19}\text{J}$$

The band gap energy of most PV cell materials ranges from 1.0 to 1.6 eV. Most solar cells can only use about 55% of the light that hits them, because some energy is below the band gap energy, and some is excess energy that cannot be used (Energy.gov).

The energy of light varies with wavelength. Light travels in waves, and one wavelength is the distance from one wave peak to the next. Colors like blue and purple have shorter wavelengths, while colors like red and orange have longer wavelengths. Light with shorter wavelengths carries more energy than those with longer wavelengths. Infrared light has about 0.5 eV, while ultraviolet has about 2.9 eV. Infrared light has a longer wavelength than red visible light, and ultraviolet light has a shorter wavelength than violet visible light. Most solar panels can only use light within the visible spectrum.

Types of Solar Cells

There are two main types of PV solar cells: crystalline and thin-film. Monocrystalline cells are made from an ingot composed of a single crystal. An ingot is a large block of metal, typically in a long cylindrical shape. These cells are usually the most efficient, but are expensive to manufacture. The wafers cut from the crystalline ingot are round, so the entire solar panel is not covered. This results in significant wasted space, leading to a larger required panel area for energy production. Polycrystalline cells are similar to monocrystalline cells, but fabricated by melting many crystals together growing from a singe crystal. They are generally less efficient than monocrystalline cells, but are cheaper to make. Advances in manufacturing technology have allowed newer polycrystalline solar cells to approach and surpass the efficiencies of monocrystalline cells. Monocrystalline cells are typically 15-20% efficient, and polycrystalline cells are usually 13-16% efficient (Energy Informative). The low manufacturing cost and increasing efficiency has caused polycrystalline cells to become increasingly popular. Silicon is the most widely used material for crystalline solar cells.

Depositing a semiconducting material onto a foundation makes thin-film cells. The semiconducting material can be deposited by co-evaporation, a method where the materials are evaporated and pumped

76

into a vacuum chamber where they grow onto the desired surface. The most widely used deposition method is known as plasma-enhanced chemical vapor deposition (PECVD). Using plasma during chemical vapor deposition can lower the temperature required for the reactions that lead to film deposition. Recent advances have led to the development of printed thin-films, which can be produced outside of a vacuum chamber. This method is less complicated and expensive than methods like PECVD (MIT Technology Review). Thin-film cells are more temperature resistant than crystalline cells, are cheaper to manufacture, and can be flexible, which increases potential applications. Cadmium telluride, amorphous silicon, copper indium gallium selenide, and gallium arsenide thin-films are used frequently in solar cells. There are other materials used for solar cells, but the ones listed here are the most widely used, and will be discussed more in depth in the following sections.

The PV effect, or the creation of electricity from exposure to light, was first found to occur in selenium. Selenium is a nonmetal that was discovered in 1817. Charles Fritts created the first thin-film PV device by compressing molten selenium between two metals; only one of which the selenium adhered to (First photovoltaic Devices). The efficiency of Fritts' PV device was only about 1%. In 2010, researchers developed a solar cell made of zinc oxide embedded with selenium. Zinc oxide is a cheap material, and the researchers at the Lawrence Berkeley National Laboratory claim that if only 9% of the zinc oxide cell foundation is selenium, the efficiency of the cell increased dramatically (Physics.org). Despite this advance, selenium has been widely replaced by silicon and other materials with higher efficiencies and lower manufacturing costs.

Crystalline silicon is the most widely used PV cell material. The crystal formation is made up of many silicon atoms connected to one another in a crystal lattice structure. Solar cells can be made of single crystal silicon, known as monocrystalline silicon, or polycrystalline silicon, which is a silicon material made of many crystals.

Monocrystalline silicon, developed by Jan Czochralski in 1918, is produced by melting high purity silicon, placing a silicon crystal 'seed' into the liquid, and pulling it slowly from the liquid to allow more silicon atoms to attach to the crystal lattice (National Solar Power Research Institute, Inc.). This method results in a large ingot made up of a single crystal. The ingot produced, which can be up to 2 m tall, is sliced into wafers a few microns thick that can be used in solar cells. The uniform structure of monocrystalline silicon increases the efficiency of the solar cell by making it easier for the electrons to flow through the silicon structure.

Polycrystalline silicon is made up of many grains of monocrystalline silicon. Metallurgical grade silicon is melted, then purified chemically or through a metallurgical process to produce polycrystalline silicon wafers for use in solar cells. Polycrystalline silicon is generally less efficient than monocrystalline silicon because it is more difficult for the moving electrons to pass from one crystal structure to the next than it is to move within one crystal. Newer technologies have been approaching monocrystalline silicon cell efficiencies, resulting in increased polycrystalline cell popularity. Although it is less efficient, there are some advantages to using polycrystalline silicon over monocrystalline silicon. Silicon wafers made of polycrystalline silicon are generally stronger and can be cut into thinner wafers. This results in more wafers per ingot.

Amorphous (non-crystalline) silicon (a-Si) cells are a type of thin-film cell. A-Si cells are used frequently to power small devices like calculators. The first thin-film cells were made of a-Si, but these cells generally have the lowest efficiency of thin-film materials used today. A-Si cells are generally comprised of a foundation layer with a silicon layer deposited onto it. In a-Si cells, silicon acts as both the n-type and p-type material. Two silicon layers are separated by a third silicon layer that acts as a junction. A-Si can be deposited at low temperatures, down to 75 °C, which allows for deposition onto plastics as well as metals. A-Si cells degrade over time due to a phenomenon known as the Staebler-Wronski Effect, reducing cell efficiency significantly in a matter of months. Within the first 1,000 hours of operation, the efficiency can decrease by 15% (Science Daily). Very thin layers of a-Si can be used to overcome this effect, but using very thin films reduces light absorbance, and therefore efficiency. To increase efficiency, layers of cells made of the thin n-type, p-type, and junction can be stacked on top of one another. A-Si films are cheap to manufacture, and can be deposited very uniformly over large surfaces (Solar Facts and Advice).

Cadmium telluride (CdTe) is a semiconductor material made of cadmium and tellurium. The main advantage to using CdTe is that very thin layers of it can absorb about 90% of the light spectrum. It is also fairly cheap and easy to manufacture. CdTe cells are made by depositing a very thin film of CdTe onto a metal panel, usually made of aluminum. The CdTe can be deposited chemically in a vacuum chamber, using a printer, by screen-printing, or by spraying (National Solar Power Research Institute, Inc.). CdTe solar cells usually consist of a layer of CdTe on top of an aluminum layer, with a cadmium sulfide (CdS) layer on top of the CdTe layer. A conducting layer, usually made of zinc oxide goes on top of the CdS layer, and this layer is topped with glass. CdTe used for solar cells is limited because CdTe cells are not as efficient as silicon based cells, and cadmium is toxic. In 2014, CdTe research efficiencies reached 20.4%, but this was still below the best performing solar based cell, which was 25.6% efficient in laboratory conditions (IEEE.org).

Copper indium gallium selenide (CIGS) solar cells have become increasingly popular. As of September 2014, commercial CIGS solar cells were the most efficient thin-film cell with 14.5% efficiency. Commercial CdTe cells were 14.4% efficient, and amorphous silicon cells were 8.1% efficient (Energy Informative). CIGS cells can be manufactured using chemical deposition like CdTe cells, printing, sputtering, or ion-beam deposition. Sputtering is the ejection of a material onto a substrate, and ion-beam deposition uses a beam of ions to carry the material to the foundation. CIGS films can be deposited onto glass, plastic, stainless steel, or aluminum, allowing for use in a wide variety of applications including flexible cells (Solar Power World). CIGS cells usually are comprised of a bottom glass, aluminum, or polymer foundation topped with a molybdenum layer, followed by the CIGS layer. A CdS layer tops the CIGS layer, and a zinc oxide layer goes over the CdS layer. Laboratory CIGS cells have reached efficiencies of 20%. Low cost and efficiency improvements are predicted to make CIGS cells increasingly popular.

Passive Solar Applications

Integrating photovoltaics (BIPV) into buildings are an application of solar cells that use the cells both for electricity generation and as an outer surface for the building. There are three types of BIPV: roofing, façade, and glazing. PV cells can replace some roof shingles, or be used as a roof for a building. One large solar panel can be used, or the roof can be composed of many small solar panels used as shingles. PV cells can also be a part of the façade of a building, or the walls. Thin cells can replace windows, or cover entire walls. Glazing refers to using semi-transparent, very thin PV cells as windows or skylights. This application is often used in greenhouses.

The amount of solar energy an area receives is known as insolation. Insolation is important to take into consideration when designing BIPV. It is usually measured in units of kilowatt-hours per square meter per day. Other factors that affect the performance of solar cells are climate, shading, and distance from the equator. Sunlight intensity varies as a function of climate or shading by near by trees and thus impacts the amount of much electricity produced from solar cells. To achieve the best cell efficiency, the sunlight should be shining directly onto the cell. At latitudes far from the equator, the sun shines down at an angle. In these locations, the panels need to be tilted to fully take advantage of the available energy. Both wafer and thin-film PV cells can be used in BIPV applications, although thin-film cells may be more easily integrated into the building structure (Solar Energy Industries Association).

Solar thermal energy systems differ from PV systems because they generate heat instead of electricity. The heat produced can be used to heat water or other heat exchange fluids and can also power solar cooling systems. Solar thermal panels are placed on roofs and contain circulating fluid that is warmed by the sun as it passes through the panels. The fluid carries the heat to a water storage tank where it raises the temperature of the water. When the hot water is turned on, the solar-heated water feeds the water heater, so less energy is needed to heat the water to the appropriate temperature. Solar thermal panels can be used to heat water for home hot water use, swimming pool or hot tub use, for heat systems, or for a combination of these (Sun Water Solar).

Another way to generate electricity from solar power is using solar heat to power a turbine. This kind of energy generation is known as concentrating solar power (CSP). The power produced using this method can be stored for use at night or when solar power production decreases due to clouds. CSP systems use mirrors to focus sunlight onto a container of heat transfer fluid. This fluid is then used to create steam, which will drive the turbine. Mirrors can also be used to focus sunlight onto PV cells to increase the amount of energy that reaches the cells. This is known as concentrating photovoltaics.

The mirrors used in CSP systems can be arranged into a parabolic trough with a tube for heat transfer fluid running parallel to it. Using this system, the heat transfer fluid can reach temperatures of about

400 °C (750 °F). Instead of the curved mirrors used in the parabolic system, flat mirrors can be arranged in a curved pattern to reduce costs. This system is known as a Compact Linear Fresnel Reflector. A third CSP system is known as the power tower. Many flat mirrors are used to focus energy onto the top of a tower, where the energy heats the heat transfer fluid to over 530 °C (1,000 °F). The flat mirrors are programmed to track the sun as it moves across the sky to maximize the amount of energy focused onto the tower. Dish engine systems use curved mirrors to create a concave surface that reflects sunlight onto a receiver located at the focal point of the mirrors. These dishes rotate to track the sun along two axes. Hydrogen is usually used as the heat transfer fluid in these systems, and can be heated to temperatures of about 650 °C (1,200 °F). In 2011, there were CSP plants in Arizona, California, and Nevada. Nine plants near Kramer Junction, CA produce over 350 megawatts of energy using parabolic mirrors. Most of the CSP plants in the United States are located in the southwestern states because direct sunlight is needed, and areas that receive little cloud cover are ideal (Solar Energy Industries Association).

The Price of Solar Power

Residential solar systems typically have 3-8 kilowatt hour (kWh) capacities. These systems can cost anywhere from $15,000-$40,000 to install, although this price can be reduced by government incentive programs. Tax credits, cash rebates, subsidized loans at low interest rates, solar renewable energy certificates (SRECs), and production based incentives (PBIs) can reduce the purchase cost of solar panels 30-50%. Of the purchase price, 30% is the cost of the panels themselves and another 20% is the cost of the 'balance of system', which includes the wiring, mounting equipment, switches, a battery system, and a solar inverter. Solar inverters change the collected electricity from direct current (DC) to alternative current (AC), which is used by home appliances and lights. Fifteen percent goes to labor, and another 15% goes to permitting and inspection fees. The final 20% of the purchase price is operational costs. Another factor that influences the amount saved by using solar energy is the amount of sunlight the cells will be receiving. Areas with many sunny days throughout the year and direct sunlight will produce more solar energy than those that have many cloudy days.

Some states require that a certain percentage of energy come from renewable sources. These states will award SRECs to homeowners for the amount of energy produced by solar panels, which can then be sold to the utility companies. These certificates only depend on the amount of energy produced, so even if a home isn't using all of the energy produced, the owner can still receive SRECs for it. States may also provide PBIs based on the energy production of a solar system. In addition to saving homeowners money on electric bills and through incentive programs, installing solar panels can increase the value of a home. In some places, homes can be fitted with net metering. Net metering allows homeowners to send energy they are producing but not using to the utility company for credits. When the solar system isn't producing enough electricity for their home, the credits are used for electricity from the utility company. This is another way to save money on electric bills and maximize the benefit of the energy produced by a system.

Businesses can also receive rebates, SRECs, and PBIs for their solar panel systems. In addition, businesses may be able to write off the value of a solar system more quickly, reducing the amount of taxes they pay on it. This is known as accelerated depreciation, and can reduce the cost of the system by 30%. Grants are available for non-profit organizations to install solar systems. These grants are typically awarded to schools, government, and towns (Energy Informative).

The cost of silicon PV cells in 2007 was around $5 per installed watt, and in 2011 it was $2.71 per installed watt. The cost of energy produced by solar cells has been decreasing rapidly, and is expected to keep decreasing due to technological advances in cell efficiencies and cell manufacturing. Businesses developing solar cells are in a very competitive market, and the most successful companies will be the ones able to minimize installation and maintenance costs, in addition to maximizing efficiency and cell lifetime. In countries where labor is very cheap, including India and China, the cost of installation can be very low. Solar power may become increasingly popular in these areas, especially in China, where air pollution from coal-fired power plants create visibility and health problems in many cities (*Energy for Future Presidents*).

Solar energy has been consistently popular according to national polls in the United States. In a poll conducted by the Solar Energy Industries Association (SEIA) in 2012, 92% of polled Americans agreed that the United States should work to develop and utilize more solar energy. The SEIA also reported that solar energy has been consistently popular in polls from 2007-2012. Additionally, 85% of voters in the same poll preferred solar over all other sources of energy in 2012. Solar power typically receives bipartisan support from Republican, Democratic, and Independent political parties as well. This widespread support of solar power contributes to increased government grants and tax breaks, further lowering the cost of buying and installing solar panels.

Although solar power is considered very clean, there are concerns about the environmental impact of solar cell production. To generate a lot of solar power, a large amount of land needs to be used for ground-mounted panels. The Department of Energy Resources (DOER) encourages the placement of solar panels in areas that do not need significant tree cutting, including industrial or vacant land. Many of the metals used in PV solar panels, especially cadmium, and some of the chemicals used in wafer production can be toxic. However, the intermediate chemicals used in panel production are carefully controlled, and the materials used in the panels themselves are contained within strong glass layers that will not break under normal conditions (Clean Energy Results).

Solar Use Today

In 2014, enough solar energy systems were installed in the United States to provide 17,500 megawatts of electricity. This is enough energy to provide electricity to over 3.5 million homes. In 2014, the number of homes and businesses in the United States outfitted with solar panels reached nearly 600,000. The price per installed watt of solar energy has dropped by over 45% since 2012, and is projected to continue to decrease (Solar Energy Industries Association). Although this seems like a lot of energy and a large price decrease, solar still provides only about 0.3% of the energy used in the United States.

One barrier to more widespread use of solar power is availability. Solar is a resource that is only available at certain times. It cannot be produced at night, and production decreases significantly when it's cloudy or when shade falls on cells at certain times during the day. Deserts in the southwestern United States are good areas for solar energy production, however the energy produced must be transported to consumers from where it is collected because not many people live in these areas. As energy is transmitted along power lines, much of it is lost (Institute for Energy Research).

As of 2015, a silicon solar panel for home or business use manufactured by Kyocera can achieve 16% efficiency. These panels can withstand up to 113 pounds per square foot of pressure from snow or ice, 51 mile per hour winds, and being hit with 1inch diameter hail. These panels are relatively low maintenance; they require wiping with a cloth or sponge to remove dirt and dust occasionally. The Kyocera panels are guaranteed to produce 80% of maximum power for 20 years.

Another solar panel manufactured by Canadian Solar is made of monocrystalline silicon, and can reach efficiencies of 15.9%. These panels can also withstand 113 pounds per square foot, and the panels are guaranteed to product 95% of the maximum power for one year, and 80% of maximum power for 25 years. These panels are advertised as self-cleaning, meaning that if the mounting angle is steep enough, rainwater will wash the panels clean.

Grape Solar, a company based in Eugene, Oregon, manufactures monocrystalline silicon solar cells that achieve 15.21% efficiency. The panels can withstand up to 50 pounds per square foot, and the materials and workmanship are guaranteed for 10 years. The panels are guaranteed to produce 90% of the 390-watt power output for 10 years, and 80% of the maximum power output for 25 years. These panels are relatively low maintenance, requiring only an occasional wipe with just water to remove any accumulated dirt (Live Science).

In 2014 SolarCity, a company based in San Mateo California, installed 7,800 panels on almost 10 acres of the Oregon Institute of Technology (OIT) campus. These panels are shown in Figure 2. Currently, the OIT campus buys the energy produced by these panels, and after 10 years of purchasing the power produced will own them. In March 2015, the panels were producing between 1.3 and 1.4 mega-watt hours per day. These purchases supplement the energy produced by the geothermal plant on the campus, which heats and provides electricity to the dorms and other campus buildings. The panels currently power about 35% of the OIT campus. The geothermal and solar plants working together provide about 100% of the electricity and heating power the campus needs, minimizing the amount of electricity bought from the utility company. Geothermal and solar power production complement each other; geothermal production is relatively stable, and can provide electricity at night when the solar panels can't. During the day, when more energy is required for heating and appliance use, the solar panels are most productive and supplement the geothermal energy production.

The arrays at OIT are about 3 by 18 ft, and are made up of several smaller panels. The panels are situated above campus, mounted on a south facing hill. There are several banks of panels, and the banks on the outer edges are smaller than those in the middle. This is to take full advantage of direct afternoon sunlight. The panels are stationary because the hill receives a lot of wind, which can damage moveable panels. During the summer, the panels produce power about 12 hours per day. In the winter, when the daylight hours decrease, the panels only produce power for about 7.5 hours each day. The angle the panels are tilted at is optimized to allow the most sunlight possible to contact them, and also makes them self-cleaning. During the year the panels have been running, SolarCity has only had to perform maintenance on the rectifiers that transform the direct current (DC) electricity to alternating current (AC), which is used in most homes for appliances. The panels installed at OIT are low maintenance, and are expected to last for 15-20 years before replacement. These panels are a great example of solar energy used on a small scale.

The Future of Solar

As concerns about atmospheric carbon dioxide levels increase, the push for energy sources with no greenhouse gas emissions grows. With widespread support from both surveyed Americans and politicians, solar energy production will continue to increase. Considerable research has gone into the development of more efficient panels, and increasing efficiencies decrease the amount of wafers needed for the same amount of energy production. This reduces the cost to the consumer, encouraging the implementation of panels on homes and businesses. In the United States, the amount of solar panels used on homes, businesses, and public buildings will continue to increase as the price for solar decreases.

For solar energy to become widely used, an alternate electricity supply must be available to supplement it during times when solar power cannot be produced. Cloud cover and snow can significantly reduce solar energy production, and no solar power can be produced at night when the sun isn't shining. Batteries are an option, but an efficient, reasonably sized battery able to store the kind of energy needed for utility companies to provide solar power to a large number of customers has not been developed. An alternative to large-scale solar production and storage is a more local approach. Like at the OIT campus, solar panels can be installed to supply energy to a smaller number of homes and businesses nearby. Additionally, a second resource can be used in conjunction with solar panels to provide a more stable source of energy. At OIT, geothermal energy is used to provide consistent power to the entire OIT campus and a nearby retirement home. Wind energy can be more productive at night in some areas, and may be a good compliment to solar energy production. Hydroelectric power production is generally stable, and is also a good option to supplement solar power. Multiple renewable resources working together will be able to provide more consistent energy to local homes and businesses, and in the future the use of a variety of renewable resources working together will be a promising way to implement wide-spread renewable resources.

Works Cited

Clean Energy Results. Ground-Mounted Solar Photovoltaic Systems, Web. 15 Mar. 2015.

Energy.gov. Passive Solar Home Design, Web. 28 Feb. 2015.

Energy Informative. Best Thin Film Solar Panels – Amorphous, Cadmium Telluride, or CIGS?, Web. 2 Mar. 2015.

Energy Sage. Solar Incentives and Rebates, Web. 2 Mar. 2015.

IEEE.org. What Makes for Better CdTe Solar Cells, Web. 1 Mar. 2015.

Institute for Energy Research. Solar, Web. 3 Mar. 2015.

Live Science. Best Solar Panels for Homes, Web. 3 Mar. 2015.

National Aeronautics and Space Administration. How do Photovoltaics Work?, Web. 1 Mar. 2015.

National Renewable Energy Laboratory. Polycrystalline Thin-Film Materials and Devices R&D, Web. 2 Mar. 2015.

National Solar Power Research Institute, Inc. Fundamentals of Photovoltaic Materials, Web. 1 Mar. 2015.

Norwegian University of Science and Technology. Solar Cells, Web. 2 Mar. 2015.

Npg asia materials. Advances in crystalline silicon solar cell technology for industrial mass production, Web. 2 Mar. 2015.

Physics.org. Selenium makes more efficient solar cells, Web. 1 Mar. 2015.

PVEducation.org. First photovoltaic Devices, Web. 1 Mar. 2015.

Science Daily. Light-induced degradation in amorphous silicon thin film solar cells, Web. 2 Mar. 2015.

Solar Cell Central. P/N Junctions and Band Gap, Web. 15 Mar. 2015.

Solar Energy Industries Association. Building-Integrated Photovoltaics, Web. 2 Mar. 2015.

Solar Facts and Advice. What is Amorphous Silicon? Why is it so Interesting Now?, Web. 2 Mar. 2015.

Solar Power World. What Are Solar Panels Made of?, Web. 2 Mar. 2015.

Sun Water Solar. Solar Thermal, Web. 2 Mar. 2015.

MIT Technology Review. Thin-Film Solar Upstart Nanosolar Slims down, Web. 15 Mar. 2015.

U. S. Department of Energy. The History of Solar, Web. 28 Feb. 2015.

U.S. Energy Information Administration. Renewable solar, Web. 23 Feb. 2015.

United States Environmental Protection Agency. Solar Energy, Web. 25 Feb. 2015.

Problems

1. Describe the different solar cell materials discussed in this section.
   a. Selenium
   b. Crystalline silicon
   c. Cadmium telluride
   d. Copper indium gallium selenide
2. A CIGS solar cell has a band gap energy of 1.5 eV. How much energy is this in Joules?
3. Calculate the efficiency of a silicon solar cell with a maximum cell power output of 1.5 W, a light input of 1000 W/m^2, and a diameter of 0.25 m. What is the power output of a panel 1 m wide and 1.5 m tall made of these cells? Assume only whole cells are used. Is this a reasonable efficiency?
4. What are the three different kinds of solar energy, and how does each produce energy?
5. A CdTe solar cell has an open circuit voltage of 3 V and a short circuit current of 5 A. How many cells are needed to produce 230 W of power, if each cell has an efficiency of 15%? Assume the cells have an area of 0.1 m^2, and use an input light of 1000 W/m^2.

Problem Solutions
1. Describe the different solar cell materials discussed in this section.
    a. Selenium: First material found to have PV properties, rarely used now, first discovered in 1817, etc.
    b. Crystalline silicon: Polysilicon and monosilicon crystals are used, monosilicon wafers are most widely used, etc.
    c. Cadmium Telluride: Thin film cells with high light absorbance, lower efficiencies, cheap to manufacture, etc.
    d. Copper indium gallium selenide: CIGS, popular/widely used, most efficient thin-film cells, easy to manufacture, etc.
2. A CIGS solar cell has a band gap energy of 1.5 eV. How much energy is this in Joules?

$$1.5 \text{ eV} \left(\frac{1.602 \times 10^{-19} \text{J}}{1 \text{ eV}} \right) = 2.4 \times 10^{-19} \text{J}$$

3. Calculate the efficiency of a silicon solar cell with a maximum cell power output of 1.5 W, a light input of 1000 W/m², and a diameter of 0.10 m. What is the power output of a panel 1 m wide and 1.5 m tall made of these cells? Assume only whole cells are used. Is this a reasonable efficiency?

$$\eta = \frac{1.5 \text{ W}}{1000 \frac{\text{W}}{\text{m}^2} \times \left(\frac{0.10 \text{ m}}{2} \right)^2 \times \pi \times 2} \times 100\% = 9.5\%$$

Yes, this is a reasonable efficiency for a solar cell. A bit low compared to today's cells.

$$\frac{1 \text{ m}}{0.10 \text{ m}} = \text{number of cells} = 10 \text{ cells wide}$$

$$\frac{1.5 \text{ m}}{0.10 \text{ m}} = \text{number of cells} = 15 \text{ cells tall}$$

$$10 * 15 = 150 \text{ cells}$$

$$P_m = 0.095 * 1000 \frac{\text{W}}{\text{m}^2} * \left(\frac{0.10 \text{ m}}{2} \right)^2 \times \pi \times 2 * 150 \text{ cells}$$
$$= 234 \text{ W}$$

4. What are the three different kinds of solar energy production and how does each produce energy?

a. Photovoltaic: semiconductors produce electricity when sunlight strikes the material, electrons are knocked from valence shells and travel to fill holes left by other electrons. These flowing electrons create current, otherwise known as electricity.
b. Solar thermal: Sunlight is focused to heat a heat transfer fluid. This fluid heats water used for heating.
c. Concentrated Solar Power: Mirrors focus sunlight onto a heat transfer fluid, which is used to create steam or vaporize a fluid which rotates a turbine, producing electricity.

5. A CdTe solar cell has an open circuit voltage of 3 V and a short circuit current of 5 A. How many cells are needed to produce 230 W of power, if each cell has an efficiency of 15%? Assume the cells have an area of 0.1 m^2.

$$P_m = 3 \text{ V} * 5 \text{ A} = 15 \frac{\text{W}}{\text{Cell}}$$

$$15 \frac{\text{W}}{\text{Cell}} * 0.15 = 2.25 \text{ W actually produced per cell}$$

$$\frac{230 \text{ W}}{2.25 \frac{\text{W}}{\text{cell}}} = 103 \text{ cells needed}$$

$$\frac{0.1 \text{ m}^2}{\text{cell}} * 103 \text{ cells} = 10.3 \text{ m}^2 \text{ needed}$$

Chapter 6. Hydropower
By Jordyn Clarke

Hydropower is often overlooked as a potential source for future renewable energy production. As the population of the world continues to grow, the energy needs of its occupants are increasing as well. Not only that, but the needs of each individual person is increasing as well. Developing nations are at the forefront of that change. Their energy needs per capita are increasing exponentially as their country develops. Moreover, their lifestyles are changing at a societal level as the developments lead to advancements in their lives that require energy to function. Residents of the United States often take it for granted that they have appliances such as refrigerators, toasters, etc. Now imagine an entire nation beginning to use toasters. Such a simple thing, but it would make a huge difference in the energy needs of the world, and that is only one appliance. As the world continues to use fossil fuels and the energy needs increase, it is evident that there is going to be a need for alternative energy sources. Due to concerns about global warming, renewable green energy sources are favorable, but those options are not perfect solutions. One of those alternatives is hydropower.

Hydropower is the energy that is produced from the movement of water. The earth's surface is just under 71% water. It surrounds us, and yet much of the resource is unused and untouched. The movement of water creates hydropower. The kinetic energy of that water is converted into electrical energy that can be transferred and used by people. In the instance of hydropower, the hot and cold bodies are the sun and the water, respectively. The bodies of water then act as the cold sink for the heat and energy that comes from the sun. The most common form of hydroelectric production comes from the spinning of turbines. These turbines are often located within a dam. The dam will most often be a structure built across a river. It must be sturdy and tall enough to block the natural flow of water. Once the water flow has been stopped, the wall must be strong enough to withstand the increasing pressure from the reservoir that will be created by the collection of water behind the dam.

Within that wall, made mainly of concrete, there are water tunnels that slope downward within the wall towards a turbine. Hence, they allow the flow of water when electricity is needed. The downward slope of the water passageways makes rivers with an elevation drop more ideal. The turbine system consists of turbine blades that spin as the water flows past. The blades are connected to a generator shaft. The shaft has magnetized poles at the top. As the shaft spins, the magnetic poles spin as well. As Faraday discovered, electrical current is created when magnetic fields are moved past conductors. So by having a conductor surrounding the magnetized rods, an electrical current is produced. Thus, the mechanical energy of the water flow and the rotation of the turbine blades have been converted into electrical energy.[63] A common form of hydroelectric is called impoundment. There is another form of hydropower that can be created using a river as well. It may not require a dam to be built, but is far less common. The diversion technique also uses turbines, but involves channeling a portion of the natural water flow through a different path. This technique is also known as run-of-river facility.[64]   Water storage is related to impoundment. This process involves storing energy for later use. When other energy sources produce more energy than the demand of the population requires, that excess electricity can be used to pump water into a reservoir. The water can be stored in that reservoir until the demand requires more energy than those other sources can produce. Then the water can be sent back down through the turbines to create electricity to meet that increased demand.

An interesting aspect about conventional dams is the ability to manipulate the water flow through the system based on consumer demand. At peaks throughout the day, more water can be released to create more energy. One of the significant disadvantages of other renewable energy sources is the lack of availability—energy production cannot be tied to energy demand; solar sources produce only during the day, wind turbines when the wind blows, or both methods can produce above demand and the excess is lost. Pumped storage offers a solution. When there is excess energy produced, no matter the source, the excess can be used to pump water into a reservoir, thus storing this potential energy. Then when there is a peak in demand, the reservoir water is released through the turbines to once again obtain energy in a useable form. This process will not be one hundred percent efficient, but it will

---

[63] H. Perlman, *Hydroelectric Power: How It Works*, WWW Document, (http://water.usgs.gov/edu/hyhowworks.html)
[64] *Types of Hydropower Plants,* WWW Document, (http://energy.gov/eere/water/types-hydropower-plants)

be more valuable than simply burning off excess energy. The energy is indeed lost, and there is nothing to gain from using that excess. In the case of pumped storage, the excess energy would have a purpose. It would be stored until it was needed to accomplish something productive. From this perspective, pumped storage hydropower has the prospect of becoming a sort of battery-like system. Pumped storage is not the only benefit to this problem; hydropower also offers the ability to increase the output at peak energy times.

Wave energy is another technique that has been researched and is beginning to be utilized. The behavior of the ocean surface contains a vast amount of potential energy. A wave can be seen as energy traveling through the water. In fact none of the water is being horizontally displaced. The water raises a certain height and then returns to the original location as a wave passes. The energy is what is traveling through the water. This energy usually originates from wind blowing across the surface of the water. As the crest of the wave become larger, the wind has more of an effect, and the wave increases in size and energy. When the wave reaches the shoreline, it interacts with the change in the sea floor, and it crashes into the beach. This is a great example of the hot and cold body that influences hydropower. The heat source is the sun, which causes the wind by having different air pressures and temperatures. The cold sink is the water. As it should be expected, there are other possible sources for wave origination. Earthquakes, underwater landslides, and adverse weather conditions are all sources of waves large enough to be classified as a tsunami. Tsunamis are large enough waves to not be affected by the sudden change in depth near the shoreline. Instead of crashing against the beach as the smaller more typical waves do, a tsunami will simply alter the sea level in that area. This allows the water to flow inland.[65]

The ocean offers a large expanse of area that can serve as potential energy source. There are multiple techniques and systems for harnessing it into electrical energy. One example of this technology is called a terminator, sometimes referred to as an overtopping terminator. This structure runs perpendicular to the direction of the waves motion.

---

[65] NOAA Ocean Explorer, *Waves are caused by energy passing through the water, causing the water to move in a circular motion,* WWW Document, (http://oceanexplorer.noaa.gov/facts/waves.html)

The middle contains a reservoir. Reaching out from each side of the reservoir are long arms that reach out forming a somewhat U-shaped barrier. The arms, reflectors, increase in height above the sea level, as they get closer to the reservoir. This means that the waves travel along the arms towards the reservoir. When the wave reaches the reservoir, it is at its maximum height and is able to crash over the top of the reservoir wall. Once contained in the reservoir, which is located slightly above sea level to give the water the maximum value of potential energy possible, it is sent through turbines back into the ocean. The turbine works much the same as the turbines built within the river dams. The movement of magnets against a conductor creates an electrical current that can be used for electricity.[66]

Attenuators are another type of device that captures wave energy. These were the first proposed and implemented devices for capture of wave energy. Scotland was the first country to construct and use these devices. Attenuators are long tube shaped structures that are parallel to the direction of wave travel. Similar to a boat, as the wave passes, parts of the attenuator rise and fall with the crest and valley of the wave. The force of the rise and fall of the attenuator is used to move another turbine and convert that mechanical energy into electrical energy.[66] Similar to the attenuator is the bulge wave device. Instead of segments that are connected together like the attenuator, the bulge wave device is a long rubber tube. The end of the tube has a turbine. It also runs parallel to the motion of the waves. As water begins to run through the tube, it creates a bulge. As it travels farther towards the turbine, the bulge size increases. Once it reaches the turbine, the water within the tube has reached its maximum potential and it runs through the turbine and back into the ocean. Once again that turbine spins to create electricity that can be run through the grid to meet the demand.[67]

Another technology is the point absorber. The top of a point absorber surface is buoyant, so that it either floats at the surface of the ocean or very near to it. An electrical converter lies below the buoyant top, and it is attached to the top by a metal rod. As the waves pass, the

---

[66] L. Guthrie, *Energy and the Environment, A Coastal Perspective,* WWW Document, (http://coastalenergyandenvironment.web.unc.edu/ocean-energy-generating-technologies/wave-energy/the-pelamis-wave-energy-converter)

[67] EMEC, *Wave Devices,* WWW Document, (http://www.emec.org.uk/marine-energy/wave-devices)

top of the device floats up and down over the crest and into the trough of the waves. This movement pulls and pushes the rod in and out of the electrical converter, creating electricity. Seeing as the top will move up and down no matter the direction of the wave, this device works in any orientation relative the flow of the waves. Also related to this is a technology type called the oscillating water column. This structure is designed to be a hollow column. Beneath the surface of the water is an opening that allows water to flow in and out of the air space within the column. The top has a turbine that spins when the air is expelled from the turbine due to the increased water level in the column. As the wave passes, the water level drops and air flows back inside. As the next crest of the wave passes, the water level rises, the air is expelled, the turbine spins to create the electricity, and the process repeats as the waves pass.[67] This is one of the many possible ways to convert waves into useable energy. The technologies highlighted here are not the only possible ways to harness the potential energy, but they encompass the main styles and techniques for wave energy conversion.

There are some things to consider when looking into wave energy as a possible source for energy production. This includes some of the technologically challenging physical requirements. The first challenge is how to connect the generator to the grid. In most cases the generator is located under water. So there are challenges in determining how to effectively move the energy produced to the main grid to be distributed. Something to consider about the wave devices is that they could be very susceptible to inclement weather conditions. Waves that surpass the capacity of the devices and changes in the direction of waves could not only make it challenging to produce energy during times of adverse weather, but it could also damage the equipment. Imagine a tsunami approaching the coastline. The likelihood of the devices surviving a tsunami is very low. For this reason, it is challenging to get investors when one event could make the equipment no longer functional and in need of repair.

The surface waves are not the only movement of water within the ocean. Water is always moving. Consider a map of the ocean currents. They curl over every inch of water. These currents are formed due to differences in water temperature, salinity, wind and differences in sea level topography. The rotation of the Earth also has an effect on the flow of water around the globe.[68] This is another great example of the hot and cold sink of hydropower. The sun creates the heat that alters the temperature of the water. Those temperature gradients create water flow. The sink that causes the heat to be lost is once again the water itself. Currents that travel large distances north and south are understood to be flowing with the temperature of the water. Near the equator the water is warmer, and thus it begins to exchange with the colder water to the north and south. Water may not be flowing exceptionally fast in these currents, but the energy potential is surprisingly large considering the speed of the water flow. Compared to wind, water flows incredibly slowly in the ocean currents. However, water is extremely dense relative to air. That density increases the energy potential. According to The Bureau of Ocean Energy Management, water flowing at 12 miles an hour can produce the same amount of energy as 112 mile an hour winds.[69]

Technology to harness the energy available in ocean currents is still very primitive by comparison to other technologies. Japan has just recently installed a new technology that is similar to kites. The Japanese were very dependent on nuclear power until Fukashima, so this is a clean energy alternative to try and supplement that dependency with another source. The devices are currently installed and in use. They are in the testing phase and are projected to remain as such until 2017. A likely possibility for another technology type that will prove to be successful is similar to that of wind turbines. In areas where the current is strong, turbines with blades that spin as the water moves past would be placed to capture the energy of the current. As the water moves from one location to the next, it will displace the blades of the turbines underwater. Similar to the turbine within the river dam, it will create an electrical current that can be hooked up to the grid and distributed for use. Other possible technology types include systems to tap into the daily tidal changes near shorelines rather than the water movement in currents.[72]

As to be expected there are some challenges to harnessing the energy of ocean currents. Similar to wave technology, there must be a way to take the energy that is produced and get it to the consumers. In

---

[68] *Ocean Currents and Climate,* WWW Document,
(http://earth.usc.edu/~stott/Catalina/Oceans.html)

97

the case of current technology, these devices will likely be farther from land than the wave devices would be. So in this case it will take a carefully engineered system to connect the infrastructure to the grid. That will be a challenge for the future technology. There are other technical challenges including marine growth, corrosion, and reliability.[69] It would be ideal to avoid marine growth on the devices. Any growth would have the potential to damage the equipment or to make it less efficient in converting energy. Corrosion would be something to look to avoid as well. Considering the entire apparatus would be under water, the materials used should be as anti-corrosive as possible. Any corrosion will lead to damage in the structure. This leads into the challenge of reliability. Especially because the system of machines will be held underwater, there are challenges surrounding the maintenance of these machines. Any time when there needs to be maintenance or repairs, it will be more expensive because of the nature of the process of making those repairs. All of these are things to consider about current technology, especially because this particular way of harnessing the energy from the ocean is still in the research stage of development. The fact that it is still in the research phase, however, allows for breakthroughs in handling these challenges. Still, these challenges are something to be kept in mind.

Hydropower was responsible for producing 2.6% of the energy that was consumed in the United States in 2013.[70] There are about 80,000 dams in the United States with uses ranging from irrigation, flood control, recreation, and others on top of those used for energy production. Of those 80,000 dams, only about 2,400 produce power.[2] That is only 3% of the structures in place. The dams that do not produce power are already built. They are already in locations with the correct water conditions for a dam, and the main structure has already been constructed. While the majority of these dams are not going to be able to make a significant dent in the growing energy needs of the world, they do have the potential to produce some energy. These could be influential on a micro grid scale. In areas surrounding these other dams, a smaller

[69] BOEM, *Ocean Wave Energy,* WWW Document, (http://www.boem.gov/Renewable-Energy-Program/Renewable-Energy-Guide/Ocean-Wave-Energy.aspx)

[70] IER, *Hydroelectric,* WWW Document, (http://instituteforenergyresearch.org/topics/encyclopedia/hydroelectric)

energy supply could possibly be enough to supply a home with energy, or a recreational building. For example, a recreational dam could be remodeled to have the water level remain constant. Any extra water height beyond that level would be released. Upon the release of that water, a turbine would spin and the energy could be collected from the release of that water. The electrical energy produced from the water that is released could be directly connected to the facilities surrounding the area. Then, not only would the energy used by those facilities be green, but also the company involved could advertise that it uses green energy. Seeing as the dam is already in place, it has the potential to fairly easily be converted into an energy producing structure. Then the energy released would have a purpose as well.

This is not the only type of unconventional dam that could be outfitted to produce energy while still fulfilling its original purpose. Consider a dam meant to control irrigation. Water that is going to be sent out to irrigate could be sent through turbines before being pumped away to the irrigation site. Then the energy produced from the water on its way to irrigate could be used to pump the water to its final location.

Currently in the United States, there are about 78,000 megawatts of hydropower are used.[71] While this only accounts for around 7% of the total energy usage in the United States, there are ways to increase this number. First, the systems already in place can be made more efficient in order to maximize energy output from the structures and equipment already in place. Second is to expand the infrastructure. The infrastructure for wave technology is still very minimal, especially relative to the dams that are already in place. Waves are everywhere in the ocean, however. Location becomes a concern mainly when the need arises to hook the generator up to the grid to transport the energy. According to the National Renewable Energy Laboratory, 250 terawatt-hours (TWh) per year are recoverable along the West Coast. The East Coast is expected to have a recoverable 160 TWh per year, and Alaska is unusually high with an expected 620 TWh recoverable each year. The United States uses approximately 4,000 TWh each year.[72] Including

---

[71] C2ES, *Hydropower,* WWW Document,
(http://www.c2es.org/technology/factsheet/hydropower)
[72] BOEM, *Ocean Current Energy,* WWW Document,
(http://www.boem.gov/Renewable- Energy-Program/Renewable-Energy-Guide/Ocean-Current-Energy.aspx)

Hawaii, wave energy could make up fewer than 28% of the energy used by Americans. The likelihood of wave energy infrastructure being maximized is low. The National Renewable Energy Laboratory gives the maximum amount of recoverable energy, which is far lower than the total potential that the ocean holds. The estimates on the amount of recoverable energy come from taking estimates on how other factors will affect the ability to lay infrastructure in some regions. For example, some coastal regions see heavy traffic due to trade routes and recreational activity. Heavy traffic would make it very challenging to have the structures in place to capture the energy because ships need to be able to get through the area. Another reason why the energy that is recoverable is not virtually unlimited comes from the environmental expectations of some areas. There are ecosystems scattered about that are more fragile than others and law protects many of these areas. Lastly, while ocean current technology is still in much of the research process, it is hard to obtain an estimate for the amount of energy that it could potentially provide. It is known, however, that the ocean currents hold a massive amount of potential energy. For example, according to the Bureau of Ocean Management, harnessing just one thousandth of the available energy in the Gulf Stream Current would provide 35% of the energy needed in Florida. So, while there is not a known number for how much energy the technology will be able to harness, that can be an indicator of the sheer amount of energy that can be found in a small segment of one current.

Once there is a basic understanding of the workings of the different types of hydropower, there must be a shift of focus towards whether or not these energy types are economically viable. The likelihood of convincing an entity to invest in a method of energy production that is not economically competitive with typical energy sources, such as coal and natural gas, is very small. Investments are about returns. So the profitability of the investment is important as well as the maintenance costs involved in keeping it up and running.

The easiest way to get an understanding of how the economics of hydropower work is to go through an example of one of the successful dams. Bonneville Dam is a very successful and well-known structure in the Northwestern United States. The total cost of the construction was $88.4 million. The cost of maintenance and operation of the dam is about $10 million each year. Before making an investment like this, companies will want to know how long it will take to get a return on their investment. The least amount of time that it could take to make that return is the most ideal. There is less risk involved in investments that are paid back sooner. The quicker that debts are paid, the sooner that money can begin to be made, and it is no longer possible to lose money on the investment. The two Bonneville powerhouses produce about five billion kilowatt-hours of electricity each year.[73] The national average for the cost of electricity is $0.1256 per kilowatt-hour.[74] How long will it take for Bonneville to pay for itself assuming the cost for maintenance and the price of electricity remains constant?

The total cost of the dam, C, can be found from:

$$C = I + Mt \tag{7.1}$$

Where, I is the initial investment, M is the maintenance cost per year, and t is the time in years.

The profits for the dam, P, can be found from

$$P = Eu \tag{7.2}$$

Where, E is the energy sold per year, and u is the unit cost of electricity. When the dam has returned its investment the cost of the dam will equal the profits, then.

$$P = C \tag{7.3}$$

Substituting Eqs 7.1 and 7.2 into Eq. 7.3:

$$Eu = I + Mt \tag{7.4}$$

Solving for t:

$$t = \frac{Eu - I}{M} \tag{7.5}$$

For this example E = 5,000,000,000 kWh, u = $0.1256, I = $88,400,000, and M = $10,000,000.

---

[73] The Economic Costs and Benefits of Bonneville Dam, WWW Document, (http://www2.kenyon.edu/projects/Dams/bec03wilsona.html)
[74] Pacific Power, *Residential Price Comparison,* WWW Document, (https://www.pacificpower.net/about/rr/rpc.html)

$$t = \frac{(5{,}000{,}000{,}000)(0.1256) - (88{,}400{,}000)}{10{,}000{,}000} = 53.96 \text{ years} \approx 54 \text{ years}$$

Based on these numbers it will take about 54 years before Bonneville will begin making money outside of paying for its initial investment and upkeep.

Using the same Bonneville Dam example, what is the cost to the producer to produce energy?

$$P = \frac{M}{E}$$

where P is the price to produce per kilowatt-hour, M is the maintenance cost in a year, and E is the energy output in a year.

The only costs involved with the production of the energy are the maintenance and upkeep costs. The costs to the producer to produce the energy in this case will be calculated assuming the investment costs have already been paid. In this case it means that this calculation is taking place 54 years after the construction of the dam.

$$P = \frac{M}{E} = \frac{\$10{,}000{,}000/yr}{5{,}000{,}000{,}000 kWh/yr} = \$0.002/kWh$$

These are two ways of doing cost benefit analysis. Hydropower is a significant investment. It takes quite a while to return its investment in profits. Using the numbers in the examples above, the company will be making 12.36 cents per kilowatt-hour that is produced once all debts have been paid from the construction of the dam. This is competitive with coal and natural gas energy production. When that is true and the only costs to the company are the maintenance and upkeep costs, each year the company will make $618 million. So it can be a risk to invest in a project as large as a dam the size of Bonneville, but once the investment costs have been paid, the payout is very high.

Economic factors are not the only things to consider when looking into outlets for alternative energy. Societal factors have big impacts on whether or not projects are put into action. The way that the public views a technology can be strong enough to determine if that technology is successful. Environmental impacts are not only important to the environment, but they also affect how the public views the quality of an energy source. One of the environmental impacts of hydroelectric power is the land use. When a river is dammed and the water is no longer allowed to flow freely, it collects on one side of the dam. Depending on the topography of the land surrounding the river in that area, the land that is covered by the collected water may vary. Flatter areas tend to have more acreage covered by the water, while areas with more hills tend to be confined to a smaller area. Initially when a dam is built, the ecosystems surrounding it are drastically changed. It treats the land much like a flood would. The habitats of animals can be harmed by this change in water level. After a while, though, the ecosystem will reach a new equilibrium. On the outlet side of the dam, the water level will have dropped significantly. Similar to the flooded region, this area will have a significant ecosystem change as well.

The aquatic ecosystems will also change. On the flooded side, the water will become stagnant. What once used to move is now stationary. It will become deeper as well. The types of plants and animals that thrive in that kind of ecosystem are different than those that thrive in the natural river system. Also on the flooded side of the dam, there will be significant sediment collection. All of the sediment, dirt and nutrients that would have continued to travel down the river will now collect in the basin of the lake that has been created. The water on lakeside is then generally very nutrient rich. This can lead to the water growing an excessive amount of algae and other waterweed species. If the excessive plant growth continues, it can crowd out other aquatic species such as fish. Following the same line of thought, the river side of the dam is no longer receiving the sediment from the upstream. It does receive some of the nutrients in the water from when the dam releases some, but it does not receive any of the heavy sediment that has settled to the bottom of the lake.[75]

Fish, notably salmon, have migration patterns. When they spawn, they travel upstream to a place where they can lay their eggs to hatch. Dams form barriers that block the travel of fish upstream to follow those natural migration patterns. Structures called fish ladders have been built to try and solve this issue. They are designed to replicate the natural challenges that a fish would face while travelling upstream. One of the common fish ladder designs involves a series of far smaller barriers. Usually still below the surface of the water, but that requires the fish to

---

[75] UCSUSA, *Environmental Impacts of Hydroelectric Power,* WWW Document, (http://www.ucsusa.org/clean_energy/our-energy-choices/renewable-energy/environmental-impacts-hydroelectric-power.html)

jump to move on to the next segment of the ladder. The bottom of the smaller barrier also has a small opening that helps to regulate the water flow and allow for another possible opportunity for fish to continue on and migrate to the next ladder section. These fish ladders help to regulate the ability for fish to continue on their normal activities.[76] Fish ladders, while far less stressful than not being able to migrate, are still quite stressful to the fish and can have negative health impacts. They are generally pretty successful for adult full-grown fish, but the smaller, less developed juvenile fish have a harder time with the challenge. There is also no guarantee that the fish will find the ladder. Many fish do, but there are also times that the fish accidently go through the turbines or do not find any way past the barrier at all.[15]

There are some other ways that the fish migration problems have tried to be solved. One example of this is the fish hatchery system. Fish hatcheries were originally meant to simply hatch as many fish as possible. This had many negative impacts on the ecosystem because often times there would be too many fish released than the natural ecosystem could support. They also began to have issues with domestication, inbreeding and natural selection against the hatchery fish. As this problem began to surface, concerns about the validity of the mission of fish hatcheries began to be questioned. Data was showing that there was not an increase in the population levels of fish in the surrounding ecosystems, which meant there was no scientific evidence supporting the fish hatcheries. Something had to change.

As fish hatcheries were put under intense review, changes began to be made that were aimed at minimizing the issues that the hatcheries were creating and to maximize the positive impact. Some of these changes included isolating the hatchery fish from breeding with the wild fish. This allowed the natural populations of fish to maintain their many genetic variations. The goal remains to keep the fish raised in hatcheries as close to the natural population, but this allows for the effects of the false environment to remain isolated from the successful natural populations. One of the techniques to be sure that this occurred was the marking of the fish that were released. This could mean a trimmed fin or

[76] Michigan Department of Natural Resources, *Fish Ladders and Weirs,* WWW Document, (http://www.michigan.gov/dnr/0,4570,7-153-10364_52259_19092- 46291- -,00.html)

other identifiable mark to make it clear what the origination of each fish was. This new focus took on a viewpoint more concerned with the welfare of the habitat and welfare of the ecosystem, rather than mass production of fish.[77] It has been proven successful; in the state of Washington alone, there are eighty-three hatcheries in place. In total, these sites release millions of fish each year. These practices are now essential for the economy of the surrounding areas and maintaining the fish populations. Depending on the environment and the goals of the hatchery, the different sites have different practices. One type is called supplementation. This type of hatchery focuses on maximizing the g fertilization and the fry survival rate. This maximizes the number of smolts, or young fish, that are released into the environment. Another type of hatchery uses the captive brood technique. This attempts to maximize the number of fish that reach maturity, and thus the fish are raised to complete maturity within the hatcheries. Each type has their purpose and the ecosystem surrounding the hatchery is analyzed to determine which type is appropriate for that specific hatchery.[78]

There are more techniques to try and minimize the impact of the dams on the fish population. Sometimes the juvenile fish are collected in a barge to be transported upstream. The barge is meant to protect the juvenile fish from the dangers of passing the dam. While it protects the fish from the dangers of fish ladders and turbines, it also creates a false migration environment. It successfully delivers the fish to their destination, but the challenging part is whether or not those fish learned the migration pattern from the trip and will return the following year. The first barge experiment failed in that area because the number of fish that returned was very small. In future experiments the barges were designed to allow the fish to excrete their own waste and pheromones into the water in which the barge travels. That way the fish leave a mark behind on the path that was travelled. The idea is to encourage them to follow that migration path again in later years. Generally, after the barges were redesigned, this was successful and is the practice that is used today.[79]

---

[77] J. Harrison, *Hatcheries,* WWW Document,
(https://www.nwcouncil.org/history/Hatcheries)
[78] WDFW, *Hatcheries,* WWW Document,
(http://wdfw.wa.gov/hatcheries/overview.html)
[79] J. Harrison, *Fish Transportation,* WWW Document,
(http://www.nwcouncil.org/history/FishTransportation)

Outside of the environmental impacts, there are other factors that affect the way that the public views the energy source. Some people have been known to find the dams unpleasant to look at. This, while seeming to be a petty problem, can have quite the impact on people who are forming opinions about a resource. The public's opinion on different energy sources serves as a large contributor to what actually is put into place. Without public support, it is challenging politically and economically to increase the production of a source. While the unsightliness may be a negative impact on peoples' views of hydropower, the lake that is created by the dam can be a source of recreation. This could include boating, swimming, fishing and other activities depending on the location and topography of the area. Possibly the most important factor towards many viewpoints is the fact that hydroelectric is a green energy source. Seeing as it is an entirely renewable resource and that it does not involve carbon emissions, it is a very green energy source. In that respect, hydroelectric is a great energy resource for the environment because it does not contribute to the Greenhouse Effect or Global Climate Change.

Hydropower has been around and is often forgotten in the sea of more complicated technological advances. However it has a lot of possibilities in its variations and types. Not only does the conventional source of hydropower have room to grow, but it has new areas to grow as well. Not only does it have an entire area to expand in the form of wave technology, but it also has current technology in the research stage of development. Both of those new areas of development have incredible potential as far as the sheer amount of power and availability that they could access. If the research is able to come up with solutions to some of the barriers in the viability of those as major resources in the energy spectrum, then they could easily become major players in the energy game.

In order for hydropower to become more of a contributor to the production of energy and meeting the demand of the people, it will have to expand. This does not necessarily mean that construction on twenty new dams needs to be started. In fact, it means that the existing infrastructure should be analyzed and considered as a possibility to be remodeled to create power producing structures in places that the dams already exist but are not currently set up for energy production. By doing this, the impact from building new structures could be avoided, and the large capital required to build such a structure could be minimized. This would also lessen the negative impact of societal views by not increasing the amount of infrastructure that is visible. Research must continue on alternative hydropower techniques. There is so much power stored inside the behavior of the ocean. Research that can help to make the energy accessible to be harnessed into electrical energy would make the ocean an incredible opportunity to expand hydropower.

All of those requirements are what would make it possible for hydropower to expand scientifically. Socially, it is important for the public to see hydropower as a positive resource. The impact that the public has on the way that change happens is often overlooked. Keeping hydropower in a positive light will help to keep it on the map as far as energy goes. Negative press about it as an energy source would make it very hard to get funding for any improvements, expansions, or research. A topic that is viewed as positive in the public eye is far more likely to be considered than something that is viewed as negative. For these reasons it is of the utmost importance that in order for hydropower to continue to be a strong contributor to the grid and for there to be the possibility of increasing production, three things should happen. First, the possibility of turning non-energy producing dams into energy producing ones would help to effectively increase the potential for production. Second, the research on alternative hydropower should continue, and third, the social light on the topic must remain positive.

Problems

1. Give brief explanation of how conventional hydropower works.

    Solution: A dam is built to block the natural flow of a river. The dam has turbines built in that can allow water to pass through and spin the turbine blades. As the blades spin, a shaft is rotated with magnetic rods on top. Those rods spin past a conducting surface and create electric current.

2. What are some of the set backs for wave technology as it stands today?

    Solution: Viability of connecting the device to the grid, maintenance costs due to location of devices, inability to withstand severe weather conditions etc.

3. A plant costs $10 million to build and $750,000 each year to maintain. Energy is sold for $0.15/kWh. How much energy does it need to produce (in kWh) every year if the company wants to be paid off in 5 years?

107

Solution: $E = \dfrac{I+Mt}{u} = \dfrac{(10{,}000{,}000)+(750{,}000)(5)}{(.15)} = 9.17 \times 10^7$ kWh/yr

4. A plant cost \$25 million to construct and has a non-profit organization invested in clean energy interested in funding the maintenance. The plant can produce 1 billion kWh every year. If energy is sold at \$0.10/kWh, how long will it take to pay off the cost to construct the dam?

Solution: $t = \dfrac{I}{Eu} = \dfrac{25{,}000{,}000}{(1{,}000{,}000{,}000)(0.10)} = 0.25$ years $= 3$ months

5. Critical Thinking: How would offering higher government tax deductions to companies that produce clean energy affect company business decisions on whether or not to invest in hydropower?

    Solution: There are many possible answers to this question. One example is as follows: Companies would have more incentive to invest in energy resources that are green. Since hydropower is a green energy source, higher tax deductions would give the company an opportunity to save money off of investing money they had already planned on spending. It would be more likely for them to invest in hydropower.

Chapter 7. Natural Gas
By Anthony Farr

Natural gas is a type of fossil fuel that is a mixture of hydrocarbon gases but the majority of this mixture is methane, which makes up around 82% of its composition[80]. The remaining percentage is composed of ethane, nitrogen, propane, carbon dioxide, butane, and pentane. Natural gas that is used in homes or power plants has been processed to remove impurities and typically is around 95% methane[1]. Energy is obtained from natural gas through its combustion. The chemical reaction for the combustion of methane is:

$$CH_4 + 2O_2 = CO_2 + 2H_2O$$

This reaction releases 890 kJ for every mole of methane[81]. As methane is the predominant component in natural gas this formula is representative of the combustion of natural gas.

Example Problem: What is the difference in the energy obtained from the combustion of methane in 200 kg of natural gas that is 95% methane compared to natural gas that is 82%?

Solution:

The molar mass of methane is: $1.008g \times 4 + 12.011g = 16.043g$

Moles of methane in 95% natural gas: $\dfrac{200kg \times .95}{.016043 \dfrac{kg}{mol}} = 11,843mol$

Moles of methane in 82% natural gas: $\dfrac{200kg \times .82}{.016043 \dfrac{kg}{mol}} = 10,223mol$

Difference in energy: $890 \dfrac{kJ}{mol} \times (11,843 - 10,223)mol = 1.44 \times 10^6 kJ$

Like other fossil fuels, natural gas is found underground. There are around 6,973 trillion cubic feet of proven natural gas reserves on

[80] "Center for Energy Economics, BEG/UT-Austin." Accessed February 22, 2015. http://www.beg.utexas.edu/energyecon/lng/LNG_introduction_07.php.
[81] "Bond Enthalpy (bond Energy)." Accessed March 19, 2015. http://www.chemguide.co.uk/physical/energetics/bondenthalpies.html.

earth[82]. This amount of natural gas contains roughly the same amount of energy as 73.5 times the total annual energy consumption of the United States. These are just the proven reserves, which are a very conservative estimate; there may be up to ten times more natural gas present than this value.

History of Natural Gas

Humans have been aware of the existence of natural gas for a very long time. Although its presence was known since ancient times, it was not utilized as an energy source until the 18th century. The first major use of natural gas was lighting. Around 1785, natural gas was used in Great Britain to light streets and houses[83]. The natural gas that was used for this purpose was obtained from coal mines. The origin of the natural gas industry in the United States was in 1821 when William Hart constructed the first natural gas well [84]. Natural gas transitioned into other uses after the creation of the Bunsen burner by Robert Bunsen in 1885[85]. However, regulation and transportation of natural gas was not very consistent until the 1940s.

Use of Natural Gas

Today, natural gas is used directly as a source of energy as well as in generating electricity. In 2013, natural gas accounted for 27% of the electricity generation within the United States[86]. In order to use natural gas as a source of energy there are several processes that must occur. Before natural gas is used within homes or to generate electricity at power plants, it must be extracted, processed and transported. The following sections will provide a detailed explanation of each process.

---

[82] "What Is the Volume of World Natural Gas Reserves? - FAQ - U.S. Energy Information Administration (EIA)." Accessed February 22, 2015. http://www.eia.gov/tools/faqs/faq.cfm?id=52&t=8.
[83] "A Brief History of Natural Gas - American Public Gas Association." Accessed March 19, 2015. http://www.apga.org/i4a/pages/index.cfm?pageid=3329.
[84] "History NaturalGas.org." Accessed March 19, 2015. http://naturalgas.org/overview/history/.
[85] "History NaturalGas.org." Accessed March 19, 2015. http://naturalgas.org/overview/history/.
[86] "What Is U.S. Electricity Generation by Energy Source? - FAQ - U.S. Energy Information Administration (EIA)." Accessed February 24, 2015. http://www.eia.gov/tools/faqs/faq.cfm?id=427&t=3.

Extraction

Natural gas is found in underground rock formations throughout the surface of the earth. It is often trapped between impermeable layers of rock along with oil and water. These rock formations are found underground, both on and off shore. Reservoirs likely to contain oil and natural gas can be identified using geological surveying techniques such as seismology. Seismology uses seismic waves caused by vibrations to map the geological formations present underground by measuring the rate at which they travel, which depends on the medium that the waves are traveling through. Through the information determined from this method and various others, companies are able to identify how likely a location is to contain a natural gas reservoir.

When a site has been identified as a potential reservoir, the drilling process is started. Old drilling techniques consisted of using a heavy bit to chip away at rock. Referred to as cable tool or percussion drilling, this method was typically used to drill to a depth of 400-500 feet and was powered by steam engines[87]. While percussion drilling is still used in some instances today, the main form of drilling is rotary drilling. As its name implies, rotary drilling relies upon a rotating bit to break through layers of rock. This process works very similarly to a household drill, except on a much larger scale. Rotary drilling is more effective than percussion drilling because it allows for continuous contact with the rock, resulting in shorter drilling time. Rotary drilling allows reservoirs to be reached thousands of feet into the ground.

Horizontal drilling is a more recently developed technique for the extraction of natural gas. This technique still utilizes rotary drilling, but provides the ability to drill horizontally as well as vertically. The combination of drilling direction produces a "J" shaped well rather than a purely vertical one. This new shape allows for the coverage of more of the area within the rock formation that is likely to contain natural gas. Also, multiple horizontal wells can be drilled off of a single vertical well. This results in a much larger area covered for the amount of drilling that is done. One vertical well with a few horizontal branches can take the place of several vertical wells. Horizontal drilling is most often combined with another method known as hydraulic fracturing.

---

[87] "Onshore Drilling NaturalGas.org." Accessed February 27, 2015. http://naturalgas.org/naturalgas/extraction-onshore/.

Hydraulic fracturing, commonly known as fracking, is the injection of a high-pressure fluid into a gas well to cause fracturing in the surrounding rock. The fluid is comprised mostly of water, but contains chemicals that reduce friction against the sides of the well resulting in less force needed to apply pressure. Acids are often used to help break up any matter that is residing in the well, and sand is included to keep fractures open after they are formed. The process of hydraulic fracturing is performed after the well has been drilled. By causing small fractures within the rock formations that contain natural gas, hydraulic fracturing allows the trapped gas to escape. While there is still applied pressure, the gas rises to the top of the well because it is less dense than the fracturing fluid. Then, when pressure is released the gas can be collected at the surface. The sand that is mixed in with the fluid can remain in the fractures that have been made in order to keep them open and allow for gas to be collected after the pressure is released. Fracking is used within older reservoirs to extend the production life by allowing access to natural gas that cannot effectively be collected by other methods. It is also essential to the collection of gas from tight shale formations.

Processing

The natural gas that is extracted from the earth is not the same as the natural gas that is used in your home or at power plants, which is defined as dry natural gas. Before natural gas can be used for these purposes it must go through processing to remove impurities. While all natural gas is predominantly methane, dry natural gas contains a higher percentage, around 95%, of methane than that extracted from underground. In order to obtain this ideal mixture, some of the impurities that need to be removed include water, sulfur, carbon dioxide, and hydrocarbons including ethane, pentane, butane and heavier crude oils[88]. Some of the processing is completed near the well, while more complicated processes are done at offsite processing plants. The first step is only for gas that is dissolved within crude oil and involves separating the gas from the oils. This is typically done either using pressure or gravity and taking advantage of the differing densities. Next, water vapor is removed either through absorption using a dehydrating

---

[88] "Processing Natural Gas NaturalGas.org." Accessed February 27, 2015. http://naturalgas.org/naturalgas/processing-ng/.

agent or through condensation[89]. The third step involves removing natural gas liquids such as butane and pentane. These liquids that are removed have their own value as sources of energy and are sold separately of the natural gas. Lastly sulfur and carbon dioxide are removed from the gas. Amine solutions are used to remove the sulfur due to their affinity for sulfur, and are able to remove up to 97% of sulfur from the natural gas[90].

Transportation

Natural gas is transported in a compressed gaseous state via pipelines or as liquefied natural gas (LNG) in trucks or shipping containers. The transportation of LNG is present only in situations in which pipelines are not viable, such as transporting overseas or within countries that cannot afford the necessary infrastructure for pipelines. This is because transporting LNG requires a lot of energy to liquefy the natural gas and maintain temperatures necessary to keep it in its liquid state. Pipelines are a more efficient means to transport natural gas, but are expensive to build. The average cost for pipeline in North America is around \$200,000 per inch diameter per mile[91].

$$\text{Total Cost} = \text{Length x Diameter x} \frac{\text{Cost}}{\text{Inch x Mile}} \qquad (8.1)$$

Pipelines can cover large distances and larger diameter pipes are necessary to supply enough gas to areas with high consumption, so costs of pipe projects can be quite large. In the United States there are around 305,000 miles of natural gas pipeline[92]. The majority of pipelines travel across state borders and the largest concentrations are located in Texas and the Northeast.

---

[89] "Processing Natural Gas NaturalGas.org." Accessed February 27, 2015. http://naturalgas.org/naturalgas/processing-ng/.

[90] "Processing Natural Gas NaturalGas.org." Accessed February 27, 2015. http://naturalgas.org/naturalgas/processing-ng/.

[91] Tubb, Rita. "2012 Pipeline Construction Report | Underground Construction," January 13, 2012. http://www.undergroundconstructionmagazine.com/2012-pipeline-construction-report.

[92] "EIA - Natural Gas Pipeline Network - Transporting Natural Gas in the United States." Accessed March 21, 2015. http://www.eia.gov/pub/oil_gas/natural_gas/analysis_publications/ngpipeline/index.html.

Power Plants

Electricity can be generated from processes involving the combustion of natural gas. Some natural gas power plants use the heat from this combustion to heat water to turn turbines with steam to generate power. Others types heat a high-pressured gas that then is used to turn turbines to generate electricity. There are three distinct types of natural gas power plants: steam generation, combustion turbine, and combined cycle.

Steam generation plants are the most similar to those that rely on coal or nuclear materials as a source of fuel. In this type of plant, natural gas is used as fuel to produce heat to boil water and generate steam. This steam is then sent through turbines that spin generators to produce electricity. The steam is then condensed back into water that can be turned back into steam. A body of water is necessary for this type of power plant to act as the cold body and to supply the water needed to produce steam. This type of power plant is not incredibly common making up only nine percent of natural gas power plants in the United States in 2012[93]. Steam generation plants have a relatively low efficiency of around 33%[94].

Combustion turbine plants do not generate steam; instead they rely on very hot compressed gas. Compressed air is sent into a combustion chamber where the combustion of natural gas is used to generate gas with high temperature and pressure. These gases contain a large amount of energy, and are sent directly through turbines to convert this energy into electricity. The motion of the gas is driven by both the temperature and pressure difference on either side of the turbine. The hotter this gas is, the higher the efficiency of the overall system. Typically systems can achieve temperatures of around 2,300 degrees Fahrenheit, but are limited to 1,500 to 1,700 degrees because of the metals used in the turbine are not able to sustain higher temperatures[95]. After electricity is produced, the gas is either released as exhaust or sent through a mechanism to recuperate some of its remaining heat energy.

---

[93] "How Many and What Kind of Power Plants Are There in the United States? - FAQ - U.S. Energy Information Administration (EIA)." Accessed March 21, 2015. http://www.eia.gov/tools/faqs/faq.cfm?id=65&t=2.
[94] "SAS Output." Accessed March 19, 2015.
http://www.eia.gov/electricity/annual/html/epa_08_02.html.
[95] "How Gas Turbine Power Plants Work | Department of Energy." Accessed March 1, 2015. http://energy.gov/fe/how-gas-turbine-power-plants-work.

Combustion turbine power plants made up the same percentage as steam generation plants at nine percent in the United States in 2012[96]. This type of plant also has a relatively low thermal efficiency of 20 to 35%[97]. However, due to their simple design, combustion turbine plants have a very quick start up time. This makes them ideal for acting as supplemental sources of power generation during peak demand.

The third type of natural gas power plant is combined cycle. Combined cycle plants utilize a combination of combustion turbine and steam generation. Natural gas is combusted to generate hot pressurized gas that is sent through a turbine like in combustion turbine plant. After this gas exits the turbine, it enters a heat exchanger that uses the thermal energy from the heated gas to boil water and generate steam. This steam is then sent through a separate turbine to generate more electricity. The steam is then condensed back into water and sent back into the heat exchanger to be converted back into steam. If viewed as a heat engine, the combustion chamber and the hot gasses act as the hot body and the water acts as the cold body. A graphic of a combined cycle natural gas plant as a heat engine is depicted in Figure 1.

Combined cycle power plants are the most prevalent, making up the remaining 82% of natural gas plants in the United States. This is due to the higher thermal efficiency that they achieve. The average efficiency of combined cycle natural gas plants in the United States in 2012 was 44.8%[98]. Thermal efficiency is the efficiency in which the power plant can convert the heat energy produced from combustion into energy in the form of electricity. The higher efficiency is achieved through the ability of the two cycles to generate more work from the same amount of inputted heat. The typical capacity of combined cycle natural gas plants is between 400-600 MWh[99]. This is enough electricity to power roughly 125,000 homes.

---

[96] "How Many and What Kind of Power Plants Are There in the United States? - FAQ - U.S. Energy Information Administration (EIA)." Accessed March 21, 2015. http://www.eia.gov/tools/faqs/faq.cfm?id=65&t=2.

[97] "How Gas Turbine Power Plants Work | Department of Energy." Accessed March 1, 2015. http://energy.gov/fe/how-gas-turbine-power-plants-work.

[98] "SAS Output." Accessed March 19, 2015. http://www.eia.gov/electricity/annual/html/epa_08_02.html.

[99] "U.S. Energy Information Administration (EIA) - Pub." Accessed March 19, 2015. http://www.eia.gov/forecasts/aeo/MT_electric.cfm.

How to calculate thermal efficiency

The equation used to find thermal efficiency of a system is:

$$\eta_C = \frac{W_C}{Q_{in}} \qquad (8.2)$$

This would be the equation used for either of the steam generation system or the simple cycle. The equation for a combined cycle follows the same format, but there are two instances of work within the system so it is modeled by:

$$\eta_{CC} = \frac{W_A + W_B}{Q_{in}} \qquad (8.3)$$

W_A represents the electricity generated by the combustion turbine and W_B represents the electricity generated by the steam generation cycle. Using the equation $W_c = Q_{in} - Q_{out}$ the total efficiency of a combined cycle can be represented by the equation is:

$$\eta_{CC} = \frac{(Q_{in} - Q_{int}) + (Q_{int} - Q_{out})}{Q_{in}} \qquad (8.4)$$

Q_{int} is the intermediate heat that is the output of the first cycle and the input for the second cycle. This equation can also be written in terms of efficiency as:

$$\eta_{CC} = \eta_A + \eta_B - \eta_A \eta_B \qquad (8.5)$$

Residential, Industrial, and Commercial Use
        The same natural gas that is used in power plants is used within residential, industrial, and commercial buildings. Within these areas the most common use of natural gas is heating. Other uses within industry include powering furnaces for the drying of paint, fueling boilers, and in chemical processes. Residential use also includes powering water heaters and use in cooking. According to the EIA, around 61% of homes in the United States use natural gas. In the United States, the combined use of these three categories is much larger than the amount used for the

generation of electricity. This sum adds up to 60% with industrial accounting for 28%, residential 19%, and commercial 13%[100]. The use of natural gas in all of these settings is dependent on an array of pipelines for distribution to each location of its use.

Use within Automobiles

Natural gas can also be used as fuel for automobiles. It can replace gasoline without major alterations to the vehicle. Its use as fuel for automobiles has increased as the price of gasoline has risen and that of natural gas has fallen. While it does not have as high of an energy density as gasoline, it is cheaper per mile. Using natural gas in cars does not come without drawbacks though. Because of its lower energy density, natural gas takes up more space per unit energy. In order for its use to be feasible, it must be compressed. It is typically compressed to 250 atmospheres, providing it with 11 kilowatts of energy per gallon[101]. This requires larger tanks within vehicles and limits the range that they can travel. Also, infrastructure is not established to support natural gas as a fuel source as readily as gasoline, so it may be necessary to purchase a compressor in order to fill up. The cost of a compressor significantly damages the economic incentive of natural gas, typically resulting in it being economically advantageous only if a single compressor can be used for multiple vehicles. Natural gas used for transportation finds its most predominant use within public transport and taxi services.

Economics

Currently natural gas is very economically appealing. According to the United States Energy Information Administration (EIA), natural gas has one of the lowest estimated levelized cost of electricity at a range of $14.3/MWh to $40.2/MWh for the different types of power plants[102]. Levelized cost of electricity is formulated by the overall

[100] "How Much Natural Gas Is Consumed in the United States? - FAQ - U.S. Energy Information Administration (EIA)." Accessed March 2, 2015. http://www.eia.gov/tools/faqs/faq.cfm?id=50&t=8.
[101] Muller, Richard A. *Energy for Future Presidents: The Science Behind the Headlines*, W. W. Norton & Company, 2012.
[102] "U.S. Energy Information Administration (EIA) - Source." Accessed March 1, 2015. http://www.eia.gov/forecasts/aeo/electricity_generation.cfm.

capital costs and the costs of producing electricity from natural gas, including fuel cost, operation and maintenance cost.

The majority of fuel costs for natural gas are associated with the cost of extraction and supply and demand. The supply of natural gas has been dramatically increased over the course of the last decade due to the increased feasibility of extracting natural gas from shale oil reserves. These sources were only made economically competitive due to the advancements in technology. The cost of natural gas fluctuates based on conditions within the market, but in 2014 the cost for natural gas used for electricity generation stayed around $4.50 per thousand cubic feet[103]. This equates to a cost of $0.44 per therm of energy provided. The cost of coal per therm of heat is lower at $0.24, and the fuel cost for coal to generate one kilowatt of electricity is $0.025. Natural gas is highly competitive with coal in this category, with an average fuel cost of $.035 per kilowatt of electricity generated[104]. The specific fuel cost per kilowatt-hour for different types of natural gas power plants can be determined by their heat rate. The heat rate is the required amount of thermal energy in order to generate one kilowatt of electricity. The fuel cost for a power plant can be found from this equation:

$$C_F = HC_N$$

(8. 6)

In the equation C_f represents fuel cost per kilowatt-hour, H represents heat rate in Btu/kWh, and C_N represents the cost of natural gas per Btu.

Capital costs for natural gas plants are comparatively low because they are smaller than coal or nuclear plants. The overnight (not including interest) capital cost per kilowatt of an advanced combined cycle is $1,024, and the same cost for an advanced combustion turbine is $676[105]. This low cost per kilowatt can be attributed to a few factors. The technology required to establish a natural gas power plant has been around for a long time, so cheap ways to produce the necessary infrastructure are present. Also, because natural gas can be integrated

---

[103] "U.S. Natural Gas Prices." Accessed March 3, 2015, http://www.eia.gov/dnav/ng/ng_pri_sum_dcu_nus_m.htm.

[104] "What Is the Efficiency of Different Types of Power Plants? - FAQ - U.S. Energy Information Administration (EIA)." Accessed March 21, 2015. http://www.eia.gov/tools/faqs/faq.cfm?id=107&t=3.

[105] "U.S. Energy Information Administration (EIA) - Source." Accessed March 3, 2015. http://www.eia.gov/forecasts/capitalcost/.

effectively with the current grid system there is no additional cost that is associated with the need of energy storage for sources like wind and solar.

The lower capital cost of a natural gas power plant also results in a faster return on investment for a power utility company. The smaller scale of natural gas power plants is appealing to investors because it poses a smaller investment risk. These qualities of natural gas power plants make them more likely to be invested in by large power utility companies. The rate of return on an investment in a natural gas power plant is determined by the capital cost of the plant, the capacity of the plant, the cost of generating one kilowatt hour of electricity, and the price that the utility can sell the electricity at. The equation for rate of return for one year is:

$$R = \frac{8,760 \times F(P-(O_F+O_v+C_F))}{C_C} \tag{8.7}$$

Where F represents the capacity factor of the plant, meaning the percentage of time the operates at full capacity; P represents the average price at which the utility can sell one kWh of electricity; O_F and O_V represents the fixed and variable operation and maintenance costs associated to the plant (including fuel cost) per kWh generated; C_F represents the fuel cost per kWh generated and C_C represents the capital cost of the plant per kWh of capacity. The constant at the front of the equation is the number of hours within a year. Capacity is not factored into this equation because it is in cost per kWh. A similar equation can be used to fine the payback period, the period in which an investment can be fully paid off, for a natural gas power plant. This equation is:

$$C = 8,760 \times F \times T(P-(O_F-O_V)) \tag{8.8}$$

Or, rearranged to find T directly:

$$T = \frac{C_C}{8,760 \times F(P-(O_F+O_V+C_F))} \tag{8.9}$$

Where T represents the payback period for the power plant and the rest of the variables are the same as in the previous equation. Because these equations do not factor in the interest that must be paid on the capital cost of the power plants the rate of return will be higher and the payback

119

period will be shorter than they would be in a real world scenario. U.S. Energy Information Administration[106] provides data and forecasts for the capital costs, operating costs and heat rate for the different types of natural gas power plants.

Example: Using data from the EIA table, calculate the rate of return and payback period for an Advanced CC (Combined Cycle) plant with a capacity factor of 87%. Assume that electricity is sold at a price of $0.12/kWh and natural gas costs $4.5/ thousand cubic feet.

Solution:

$$C_C = \frac{\$1,023}{kWh} \quad F = .87 \quad P = \frac{\$0.12}{kWh} \quad O_F = \frac{\$15.37}{kWyr} \quad O_V = \frac{\$3.27}{MWh} \quad C_F = \frac{\$4.5}{Mcf}$$

Convert O_F to $/kWh: $O_F = \frac{\$15.37}{kWyr} \times \frac{kWyr}{8760kWh} = \frac{\$0.00175}{kWh}$

Convert O_V to $/kWh: $O_V = \frac{\$3.27}{MWh} \times \frac{MWh}{1000kWh} = \frac{\$0.00327}{kWh}$

Calculate C_F in$/kWh: $C_F = \frac{\$4.5}{Mcf} \times \frac{1Mcf}{1,025,000Btu} \times \frac{6430btu}{kWh} = \frac{\$0.0282}{kWh}$

Calculate rate of return: $R = \dfrac{8760 \times F(P - (O_F + O_v + C_F))}{C_C}$

$$R = \frac{8,760 \times .87(\$0.12 - (\$0.00175 + \$0.00327 + \$0.0282))}{\$1,023} = 64.6\%$$

Calculate payback period: $T = \dfrac{C_C}{8760 \times F(P - (O_F + O_V + C_F))}$

---

[106] "U.S. Energy Information Administration (EIA) - Pub." Accessed March 19, 2015. http://www.eia.gov/forecasts/aeo/MT_electric.cfm.

$$T = \frac{\$1,023}{8,760 \times .87(\$0.12 - (\$0.00175 + \$0.00327 + \$0.0282))} = 1.55\,yr$$

Societal Factors

The economic advantage of natural gas stems from a combination of low capital, operational and maintenance, and fuel costs. Of these costs, the cost of fuel is the most volatile, susceptible to changes in supply and demand in the world market. As long as fuel costs remain low, natural gas will remain the one of the most economically competitive sources of energy.

Sustainability

When looking at any source of energy, its sustainability should be a factor that is taken into consideration. Questions that could be raised regarding natural gas include the fact that it is a finite resource, its contribution to global warming, and the environmental impacts of its extraction and processing.

Natural gas is a finite resource because it is being consumed more rapidly than it is being replenished through natural processes. It is a fossil fuel that has been formed by the decomposition of organic matter that undergoes chemical reactions due to heat and pressure from the earth. The depletion of this resource is not a major concern today due to the technological developments in extraction methods that have drastically increased the viable reservoirs. However, the availability of natural gas will ultimately cease to exist. This fact should be taken into consideration when developing the methods and sources of energy that are used to support the growing need for electricity throughout the world. In the long run it will be better to invest in an energy source that is sustainable and will continue to produce over time. On top of this very distant concern, there are also more pressing concerns about natural gas as an energy source.

As natural gas is comprised primarily of methane and its combustion results in the emission of carbon dioxide into the atmosphere, concerns about climate are relevant to its use as an energy source. Both of these gases are known as greenhouse gases that trap heat within the atmosphere and contribute to global warming. Carbon dioxide is also released from other sources of energy such as coal. Natural gas electricity generation has been shown to produce around half the amount of carbon dioxide emissions per amount of energy produced when

121

compared to coal powered plants[107]. This makes natural gas the lesser of two evils in carbon dioxide emissions and the preferred alternative to coal. The amount of CO_2 emitted by the amount of natural gas required to generate one kilowatt of electricity is 1.22 pounds compared to the 2.16 pounds emitted by coal[108]. Natural gas also does not emit sulfur compounds, mercury or other noxious pollutants that coal power plants do. There is also concern about the methane that is emitted during the extraction of natural gas. Methane's effect as a greenhouse gas is over 20 times greater than that of an equivalent amount of carbon dioxide[109]. Small amounts of methane are leaked into the atmosphere during every process of natural gas production and use. Conclusive research has not been done to measure the scale of this leakage for the entire natural gas industry, but more research is likely to be done because this issue is very relevant to concerns about the contribution of the natural gas industry to global warming. Currently natural gas seems to be a better alternative in regards to carbon emissions and climate change. However, it is not the long-term solution to ending climate change, especially if methane leakage turns out to be higher than it is currently thought to be. It may serve as a bridge towards a cleaner energy portfolio, especially in developing nations like China, but it still negatively contributes to climate change when compared to energy sources such as wind, solar, or hydro.

Practices associated to the extraction of natural gas have also had a history of negative environmental impacts. Hydraulic fracturing has in some instances contaminating sources of drinking water with chemicals used in the process. This contamination can occur from the unintended fractures between the well and groundwater. This results in the leakage of the fluids from the fracking process into the groundwater. In a town in Pennsylvania near a well that was undergoing hydraulic fracturing, residents experienced ignitable tap water after natural gas had leaked into the ground water due to the fractures made around the well. There

---

[107] "How Much Carbon Dioxide Is Produced per Kilowatthour When Generating Electricity with Fossil Fuels? - FAQ - U.S. Energy Information Administration (EIA)." Accessed March 1, 2015. http://www.eia.gov/tools/faqs/faq.cfm?id=74&t=11.
[108] "How Much Carbon Dioxide Is Produced per Kilowatthour When Generating Electricity with Fossil Fuels? - FAQ - U.S. Energy Information Administration (EIA)." Accessed March 20, 2015. http://www.eia.gov/tools/faqs/faq.cfm?id=74&t=11.
[109] "Methane Emissions | Climate Change | US EPA." Accessed March 1, 2015. http://epa.gov/climatechange/ghgemissions/gases/ch4.html.

is also concern about the proper storage and management of the chemicals that are used for hydraulic fracturing. Hydraulic fracturing requires millions of gallons of water for each time a well is fracked. This results in very large quantities of wastewater that must be stored. Typical storage techniques consist of open-air ponds or keeping the water in injection wells. If the wastewater leaks into the environment it can cause serious damage to the surrounding ecosystems. There is also concern about hydraulic fracturing resulting induced earthquakes.

Grid Compatibility

Natural gas as a source of electricity fits within the current grid infrastructure better than any other source. Unlike energy sources such as wind or solar, it provides a constant output of electricity. This constant output can also be ramped up or down to fit the needs of the grid. Also combustion turbine natural gas power plants have the quickest production start up of any form of major electricity generation. This makes this type of plant ideal for providing immediate backup when the grid experiences a spike during peak demand. All of these qualities of natural gas allow it to be easily integrated within the current grid system.

Future of Natural Gas

As it appears today, natural gas is guaranteed to play a very large role in the energy market over the next 50 to 100 years. Increased access to the reservoirs on earth due to the viability of extraction from shale oil due to horizontal drilling and hydraulic fracturing has secured an important role for natural gas. This efficient means of extraction and the relatively low capital costs associated with natural gas plants will allow for natural gas to outcompete other sources of energy from an economic standing. Also, because natural gas produces less carbon dioxide emissions than coal, it will benefit from the concerns about climate change. It is predicted that natural gas will make up 73% of capacity additions between 2013 and 2040[110].

The future of natural gas beyond a hundred years is more difficult to gauge. This is in part due to the fact that it is difficult to predict ideals that will be valued by society that far into the future. However, some attributes about natural gas ensure its questionable

---

[110] "U.S. Energy Information Administration (EIA) - Pub." Accessed March 19, 2015. http://www.eia.gov/forecasts/aeo/MT_electric.cfm.

future in the very long term. Some of these issues are that the reserves of natural gas are finite, and the combustion of natural gas contributes to global warming.

As noted at the beginning of the chapter, the world reserves of natural gas are proven to be 6,973 trillion cubic feet. In order for reserves to be classified as proven, they must "demonstrate with reasonable certainty to be recoverable in future years from known reservoirs under existing economic and operating conditions.[111]" This estimate is relatively conservative and the amount of proven reserves has steadily grown every year for the past decade. While reserves currently appear abundant, they will eventually run out. This would mean an end to natural gas as a source of energy.

However, there is another known source of natural gas that is not currently a technologically or economically feasible source, but could be developed into one in future circumstances. This source is methane hydrates, a combination of methane gas and water with a chemical formula of $CH_4.6H_2O$. Methane hydrates form under conditions with low temperature and high pressure. Methane hydrates are found on the ocean floor and throughout permafrost. It is predicted that methane hydrates contain more than double the amount of carbon contained in the sum of all fossil fuel reserves[112]. This makes methane hydrates a massive potential source of energy. If a method of safe and cost effective collection of methane from this source were to be developed it could greatly extend the lifetime of the natural gas industry. Methane can also be obtained through biogas systems that collect methane from the decomposition of garbage and other organic materials, but this falls under biomass as an energy source.

The emission of carbon dioxide is another concern that will influence the future use of natural gas. While it is cleaner than coal, it still emits a sizeable amount of carbon dioxide. Depending on how climate change develops over the next century these emissions could

[111] "Proven Reserves: An Estimated Quantity of All Hydrocarbons Statistically Defined as Crude Oil or Natural Gas, Which Geological." Accessed March 19, 2015. http://www.opec.org/library/Annual%20Statistical%20Bulletin/interactive/2004/FileZ/definition.htm.
[112] "Cambridge Journals Online - MRS Bulletin - Fulltext - Methane Hydrates: An Abundance of Clean Energy?" Accessed March 2, 2015. http://journals.cambridge.org/action/displayFulltext?type=1&fid=7960258&jid=MRS&volumeId=33&issueId=04&aid=7960256.

determine the fate of natural gas. Measures that could be taken to help prevent this issue include carbon scrubbing and sequestration. Carbon scrubbing involves the chemical removal of carbon dioxide from emissions. While current methods of carbon scrubbing exist, they do not effectively capture enough of the carbon dioxide emitted to reduce the impacts of burning fossil fuels. However, if a method for effectively removing this greenhouse gas from emissions were to be developed, the gas could be collected and sequestered. Sequestration refers to the long-term storage of a material. Methods for sequestering carbon dioxide are being researched and the common theme is that underground storage is likely to be an effective option. Underground formations such as old oil or gas reserves, or sedimentary cave systems are proposed storage locations. If carbon dioxide is able to be removed as a byproduct of electricity generation from natural gas and is able to be safely stored instead of residing in the atmosphere, the future of natural gas will not be dictated by concerns about climate change.

Problems

1. How much would a pipeline with a diameter of 20 inches stretching from Portland to Seattle cost if the average cost is $200,000 per inch per mile?

Solution:

Use equation: $Total\ Cost = Length \times Diameter \times \dfrac{Cost}{Inch \times Mile}$

$Total\ Cost = (145)(20)(200,00) = \$580\ million$

2. If the combustion of a natural gas in a combine cycle power plant generates 1.8×10^8 Btu of heat energy per hour, the combustion turbine cycle has a thermal efficiency of 31%, and the steam cycle produces 1,200 kWh of electricity, what is the efficiency of the steam cycle? What is the overall efficiency?

Solution:

Convert Btu to kWh: $1.80 \times 10^7\ Btu \times 0.000293 \dfrac{kWh}{Btu} = 5,274 kWh$

$$Q_{in} = 5,274 kWh \qquad W_B = 1,200 kWh \qquad \eta_A = .31$$

Find W_A: $\eta_A = \dfrac{W_A}{Q_{in}} \rightarrow W_A = \eta_A Q_{in} = (.31)(5,274) = 1,635 kWh$

Find Q_{int} : $W_A = Q_{in} - Q_{int} \rightarrow Q_{int} = Q_{in} - W_A = 5,274 - 1,635 = 3,639 kWh$

Find η_B: $\eta_B = \dfrac{W_B}{Q_{int}} = \dfrac{1,200}{3,639} = 33\%$

Find η_{CC}: $\eta_{CC} = \eta_A + \eta_B - \eta_A \eta_B = (.31) + (.33) - (.31)(.33) = 54\%$

3. How much methane would have to be released in the generation of 330 MWh of electricity to equal the greenhouse effect of the carbon dioxide emitted?

Solution:

Convert MWh to kWh: $330 MWh \times \dfrac{1,000 kWh}{MWh} = 330,000 kWh$

Amount of CO_2 emitted: $330,000 kWh \times \dfrac{1.22\,lb\ CO_2}{kWh} = 402,600\,lb\ CO_2$

Amount of methane to equal greenhouse effect?

4. What is the rate of return and payback period for a Conventional CT (combustion/gas turbine) natural gas power plant with a capacity factor of 85%? Assume that electricity is sold at a price of \$0.12/kWh and natural gas costs \$4.5/ thousand cubic feet.

Solution:

$$C_C = \frac{\$973}{kWh} \quad F = .85 \quad P = \frac{\$0.12}{kWh} \quad O_F = \frac{\$7.34}{kWyr} \quad O_V = \frac{\$15.45}{MWh} \quad C_F = \frac{\$4.5}{Mcf}$$

Convert O_F to \$/kWh: $O_F = \dfrac{\$7.34}{kWyr} \times \dfrac{kWyr}{8760 kWh} = \dfrac{\$0.000838}{kWh}$

Convert O_V to \$/kWh: $O_V = \dfrac{\$15.45}{MWh} \times \dfrac{MWh}{1000 kWh} = \dfrac{\$0.0155}{kWh}$

Calculate C_F in \$/kWh:

$$C_F = \frac{\$4.5}{Mcf} \times \frac{1 Mcf}{1,025,000 Btu} \times \frac{10850 btu}{kWh} = \frac{\$0.0476}{kWh}$$

Calculate rate of return:

$$R = \frac{8,760 \times F(P - (O_F + O_v + C_F)}{C_C}$$

$$R = \frac{8,760 \times .85(\$0.12 - (\$0.000838 + \$0.0155 + \$0.0476)}{\$973} = 42.9\%$$

Calculate payback period:

$$T = \frac{C_C}{8,760 \times F(P - (O_F + O_V + C_F))}$$

$$T = \frac{\$973}{8,760 \times .85(\$0.12 - (\$0.000838 + \$0.0155 + \$0.0476))} = 2.33yr$$

5. What percentage of electricity generation would have to be changed from coal to natural gas in order to reduce carbon dioxide emissions by 30%?

Solution:

Assume total electricity generation to be 100 kWh: $kWh_T = 100kWh$

Calculate original CO_2 emissions: $lb_1 = 100kWh \times \frac{2.16lb}{kWh} = 216lb$

Calculate the amount for 60% emissions:

$$lb_2 = lb_1 \times (1-.30) = 216lb \times (1-.30) = 151.2lb$$

Pounds contributed by each source:

$$lb_{NG} = 1.22\frac{lb}{kWh}kWh_{NG} \quad lb_C = 2.16\frac{lb}{kWh}kWh_C$$

Equation for total emissions:

$$lb_2 = lb_{NG} + lb_C \rightarrow 151.2lb = 1.22kWh_{NG} + 2.16kWh_C$$

Equation for total electricity generation:

$$kWh_T = kWh_{NG} + kWh_C \rightarrow 100kWh = kWh_{NG} + kWh_C$$

Solving the system of equations yields:

$$kWh_{NG} = 68.94kWh$$

Percentage change to natural gas:

$$\frac{kWh_{NG}}{100} = 68.94\%$$

Chapter 8. Wind Energy
By: Musa A. Moussaoui

Mankind has harvested wind power ever since a sail was attached to a primitive raft. Today, with the necessary technological innovations, people have been able to convert this true force of nature into useful electricity for everyday use. Compared to most other energy sources, wind is an easy and abundant resource to harness. Essentially, wind turbines take the mechanical energy from flowing air and transform it into functional electrical energy.

Wind

Starting at the source, air molecules, not visible to the naked eye, vibrate and collide with each other and their surroundings. By doing so, molecules of nitrogen, oxygen, and carbon dioxide are producing a force.[113] Air pressure is defined as this force over a given area. Changes in air pressure across a horizontal distance produce wind. Similar to a deflating balloon, air will naturally rush from an area of high pressure to an area of low pressure. Anyone who has seen a weather broadcast knows that this gradient of air pressure is not constant by any means. If it were, air would reach pressure equilibrium and the wind would be no more.

At its core, wind energy is a type of solar energy, like many other energy sources. Temperature is the main factor that keeps a lively diversity of air pressure across the earth. The sun is ultimately the heat source of this energy cycle. Approximately 2% of the thermal energy that the terrain of the earth—the ultimate cold sink—receives is converted into wind energy.[114] Hot air is less dense and rises due to buoyancy, while air pressure at the surface drops. This allows for cooler air to move into the lower pressure area. This is why low-pressure systems are often associated with bad weather. Furthermore, the variability in the earth's surface, shape, and tilt cause the regions near the equator to heat up the most. Conversely, the poles receive little light and heat from the sun. This difference in pressure is the main driving

---

[113] "Where Does Wind Come From",
 http://www.scientificamerican.com/article/where-does-wind-come-from/, accessed 2/7/2015
[114] "What Causes Wind",
 http://www.iowaenergycenter.org/wind-energy-manual/wind-and-wind-power/what-causes-wind/, accessed 2/7/2015

force behind most of the global wind patterns. Finally, the rotation of the earth skews wind currents to turn clockwise in the southern hemisphere and counter clockwise in the northern hemisphere.[115] This phenomenon is better known as the Coriolis Effect, which only affects large bodies and is often misattributed to draining toilets. The culmination of these effects and other factors produce a wide range of wind speeds across the earth that simultaneously enables and hinders producing power with wind.

The History of Wind Power

Early recorded history shows that people have been using wind energy as early as 5,000 B.C. in order to propel boats up and down the Nile River.[116] Classical windmills were first used in Persia to pump water and grind grain between 500 and 900 B.C.[116] Windmill use stayed relatively localized in the Middle East; until around 1,000 A.D., it spread to European countries in the north.[116] For example, the Netherlands used modified windmills to drain lakes and marshes in the Rhine River Delta.[116]

The 19th century was a time for many innovations that led to modern turbines. With the advent of electricity, wind in North America was now used to generate power for homes and businesses.[116] Developments in steel production, allowed for more efficient steel blades.[116] By the end of the 19th century, there were more than six million windmills across the countryside, as homesteaders moved west.[116] At the turn of the century, larger windmills, now called wind turbines, were erected throughout the hills of Denmark.[116] Fast forward to the time of World War II, the then-largest turbine went online, feeding Vermont's local utility network a generous 1.25 megawatts (MW).[116] In the 1950s, although wind turbines were becoming reasonably efficient, most of the wind turbines in the United States were shut down, because of disuse and competing energy sources.[116]

However, in the 1970s, due to an energy crisis and subsequent oil price boom, there was a surge in interest and research in alternative energy sources including wind.[116] In 1978, Congress enacted the Public Utility Regulatory Policies Act of 1978, requiring companies to use a

---

[115] "The Coriolis Effect",
https://www.youtube.com/watch?v=i2mec3vgeaI, accessed 2/7/2015
[116] "History of Wind Energy",
http://energy.gov/eere/wind/history-wind-energy, accessed 2/8/2015

specific percentage of electricity generated by renewables like wind.[116] In 1981, NASA scientists Larry Viterna and Bob Corrigan developed the Viterna Method to model wind turbine performance, specifically how wind speed and power were related, which even today continues to improve turbine design and operation.[117]

The United States has taken many initiatives to increase wind power generation in the recent years. Through 1990, more than 2,200 MW of wind power capacity were established throughout California, equivalent to more than half of the world's capacity for wind power at the time.[116] In 1992, the Energy Policy Act facilitated an energy production tax credit of 1.5 cents per kilowatt-hour (kWh) on wind-power-generated electricity. In just four years, from 2000 to 2004, the cost of wind-power-generated electricity dropped from 4-6 cents per kWh to 3-4 cents per kWh.[116] As of 2012, wind energy is the number-one source of renewable electricity in the United States, currently able to power 18 million homes.[118]

Wind Turbines

Aside from newer experimental designs, the prototypical three-bladed horizontal axis turbine is the most common device to harness wind energy for power generation.[119] A wind turbine can be thought as the opposite of a giant fan. In other words, fans produce wind by using electricity while turbines use wind to produce electricity. Furthermore, parallel components of fans can be found in wind turbines. First, the blades, typically three but sometimes two, are molded to maximize the wind that propels them. Engineers consider many parameters when designing these blades, because they are arguably the most important feature of the machine when optimizing the efficiency of the turbine. Wind turbine blades vary in size from 34 to 55 meters, and they are generally made out of composites like laminated woods and carbon fiber

---

[117] " Wind Energy Research Reaps Rewards",
http://www.nasa.gov/vision/earth/technologies/wind_turbines.html, accessed 2/8/15
[118] "What 2014 Meant For Wind In 5 Graphics"
http://aweablog.org/blog/post/what-2014-meant-for-wind-in-5-graphics,
accessed 2/9/2015
[119]"How Do Wind Turbines Work"
http://energy.gov/eere/wind/how-do-wind-turbines-work, accessed 2/11/2015

in order to be strong but still light.[120] The cross section of a turbine's blade resembles a flattened teardrop shape in order to achieve lift the same way that conventional airplane wings or propellers do. In particular, lift is achieved when the asymmetric shape of the wing causes air to travel slower on the upper surface of the wing, which generates an area of lower pressure. Bernoulli's principle can be simplified demonstrating that a difference in fluid velocity causes a pressure difference.[121]

$$\Delta P = \frac{1}{2}\rho(v_f^2 - v_i^2) \qquad (9.1)$$

Finally, the pressure difference over the area of the wing creates an upward lift force. The fluid friction or drag of the air passing around the wing is another important force to consider when designing a turbine's blade. Like lift, drag is also dependent on the shape and orientation of the blade. Moreover, the twist or pitch of the blade—measured as the angle of attack—can change the total aerodynamic force acting on the blade. For instance, turbine blades are manufactured with a slight overall twist to them, because the outer portion of the blade experiences relatively higher wind speeds.[122]

The three blades are all connected to a rotor, which bolts to a cast iron hub, one of the heaviest parts of the turbine at 8 to 10 tons for a 2-MW turbine.[120] The rotor is attached to the low-speed shaft, which transfers its torque to the high-speed shaft through the gearbox. This transfer of power is necessary because wind turbines turn at about 20 revolutions per minute (rpm), or once every three seconds, but electric generators require 1,200 to 1,800 rpm to run efficently.[120] However, a few wind turbines have switched out the industry standard of induction generators to permanent magnet generators that can forego the gearbox and operate at much lower speeds.[120]

Everything aside from the rotor and tower is housed in the nacelle, a large room-sized box that sits atop the turbine's tower. In order to adapt to changing winds, turbines are fitted with a dynamic

---

[120] "Anatomy of a Wind Turbine"
http://www.awea.org/Resources/Content.aspx?ItemNumber=5083, Accessed 2/14/2015
[121] "Bernoulli's Equation",
 http://hyperphysics.phy-astr.gsu.edu/hbase/pber.html, Accessed 2/14/2015
[122] "Wind Turbine Design",
https://www.youtube.com/watch?v=p5k2LhKBSgQ, Accessed 2/14/2015

control system. It receives information about wind speed and direction from the anemometer and the wind vane respectively.[120] Additionally, it reads the levels of electrical power generation, the rotor speed, the blades' pitch angle, vibration levels, and a myriad of other parameters.[120] By collecting this data, the control system is now adept to rotate the nacelle and the rotor to face the wind with the yaw drive.[120] The individual blades can also be rotated through the pitch drive to decrease the amount of lift when wind speed conditions become too great.[120]

Lastly, at 80 to 100 meters tall, the tower stands, on average, slightly higher than the overall diameter of the blades' sweep.[120] The overall sizes of turbines have been rapidly growing since the 1980s; current estimates predict that future turbines could eclipse most skyscrapers in overall height.[123] Further, larger blades can catch more air, which translates to more power. Ultimately, a turbine's power output depends on the mass flux of air passing through its sweep. This can be characterized with equation:[124]

$$P = \frac{1}{2}\rho A v^3 * C \tag{9.2}$$

Where P is the electric power extracted from the turbine, ρ is the density of the air, A is the area swept by the blades, v is the air velocity, and lastly, C is a power coefficient depending on the efficiency of the turbine—currently the upper limit or Betz Limit is 59% for a conventional wind turbine.[124] Notably, power is directly proportional to area, so the aforementioned increase in blade length holds. On top of that, the power output is cubically related to the air velocity. Hence, turbines are becoming taller in order to catch higher altitude winds that are faster and importantly, more consistent.

Example Problem: Calculate at what air velocity would a turbine need to experience to generate 1 MW electric assume a 40 meter blade, the Betz limit, and 1.2 kg per meter cubed for air density?

[123] "Wind 2013 Road Map",
http://www.iea.org/publications/freepublications/publication/Wind_2013_Roadmap.pdf, Accessed 2/15/2015
[124] "Wind Power Fundamentals",
http://web.mit.edu/windenergy/windweek/Presentations/Wind%20Energy%20101.pdf, Accessed 2/15/2015

Answer:
Start with equation 9.2:

$$P = \frac{1}{2}\rho Av^3 * C$$

Rearrange to solve for velocity:

$$v = \sqrt[3]{\frac{2P}{\rho A * C}} \qquad (9.3)$$

Substitute the know values and solve equation for velocity:

$$v = \sqrt[3]{\frac{2 * (1,000,000)\text{W}}{1.2\frac{\text{kg}}{\text{m}^3}(\pi(40\text{ m})^2) * 0.59}} = 8.25\frac{\text{m}}{\text{s}} \qquad (9.4)$$

Economics

In the United States, current utility-size wind turbines range from 100 kW to 3.6 MW.[120] The overall average capacity of all US wind turbines is about 2 MW.[120] That being the case, ostensibly, it would take only two wind turbines to power a city like Corvallis, Oregon, which annually consumes about 32 million kWh or 3.65 megawatt-years (MW-yr).[125] However, wind, like many other renewables, suffers from a significant amount of down time in power generation. Even in an ideal turbine placement like the shores of Denmark, fluctuating wind will cause variability in power generation. Capacity factor is a measure of how often an electric generator runs for a specific period of time, and it is defined as the ratio of the actual output of a generator to its maximum output. For example, if a 10-MW power plant produced 7 MW-y in a year it would have a capacity factor of 70%, which is near the average of conventional power plants.[126] The deficit can come from refueling the source, like in a nuclear, coal-fired or natural gas power plant. Nonetheless, the capacity factor of hydroelectric is still low at 40% even

[125] "Community Profile",
http://www.epa.gov/greenpower/communities/communities/corvallisorcommunity.htm, Accessed 2/16/2015
[126] "What is a Capacity Factor",
http://www.eia.gov/tools/faqs/faq.cfm?id=187&t=3, Accessed 2/17/2015

though the source of power—dammed bodies of water—is relatively constant, because the capacity factor is still compared to the maximum capacity of the generator at peak supply and demand.[127]

All of these values are essential when weighing the value of a specific resource. Most energy sources are assessed by their Levelized Cost of Energy (LCOE).[128] The LCOE is a metric in terms of cost per energy, like cents per kilowatt-hour, most commonly seen on a residential electric bill. At its fundamentals, it divides the total net cost of a power plant over its lifetime by the total energy produced by that plant. The National Renewable Energy Laboratory has a simplified equation to calculate the LCOE for any source, listed below as equation:[128]

$$sLCOE = \frac{OCC * CRF + O\&M_f}{8760 * CF} + O\&M_v + FC * HR \qquad (9.5)$$

OCC is the overnight capital cost in dollars per kilowatt ($/kW). CRF is the capital recovery factor per year, which related to annuity. CF is simply the capacity factor of source. The two O&M's are the fixed and variable operation and maintenance costs in dollars per kilowatt-year ($/kW-y) and dollars per kilowatt-hour ($/kWh), respectively. FC is fuel cost, expressed in dollars per million British thermal units ($/MMBtu). HR is heat rate, measured in British thermal units per kilowatt-hour (Btu/kWh). All of these vary with the time and collection of the dataset.

Example Problem: Calculate the LCOE of onshore wind given: OCC is 2,000 $/kW; CRF, 10% per year, O&M fixed at 30 $/kW-y, CF is 30% assuming no variation in O&M costs and fuel is free.[129]

Solution: Start with Eq. 9.5 and solve with the given parameters and assumptions.

---

[127] "Capacity Factors for Utility Scale Generators Not Primarily Using Fossil Fuels", http://www.eia.gov/electricity/monthly/epm_table_grapher.cfm?t=epmt_6_07_b, Accessed 2/18/2015

[128] "Simple Levelized Cost of Energy (LCOE) Calculator Documentation", http://www.nrel.gov/analysis/tech_lcoe_documentation.html, Accessed 2/22/2015

[129] "Cost and Performance Assumptions for Modeling Electricity Generation Technologies", http://www.nrel.gov/docs/fy11osti/48595.pdf, Accessed 2/22/2015

$$\text{sLCOE} = \cfrac{2{,}000\,\frac{\$}{kW} * \frac{0.10}{yr} + 30\,\frac{\$}{kW*yr}}{8{,}760\,\frac{hr}{yr} * (0.30)} + \cancel{O\&M_v} + \cancel{FC}$$

(9. 6)

$$\cancel{* HR}$$

$$\text{sLCOE} = 0.0875\,\frac{\$}{kWh}$$

$$\text{sLCOE} = 8.75\,\frac{cents}{kWh}$$

Offshore

 Wind power overall has a capacity factor of about 30%, but over the last few years it has been steadily increasing with innovations in turbine technologies.[127] Using offshore wind power has been of particular interest, because offshore wind speeds are stronger and can provide power more reliably with a capacity factor of 40%. [130] Furthermore, offshore wind is more desirable, because turbines can be built without obstructing scenery in the countryside, where they would be built otherwise. Nevertheless, oceans and seas also present more hostile dangers, like rough waters. However, in the event of a catastrophic failure, offshore wind turbines are safer since they are more remote from human populations and farm animals. The average distance of an offshore wind turbine farm in Europe is about 18 miles from land.[131] The biggest barriers for offshore wind are upfront investments that are significantly higher due to the expensive process of establishing tower foundations, high voltage cables, and offshore electric substations. To mitigate these costs, most countries and private companies are employing an economy of scale and building massive wind farms offshore, which is more cost effective since fixed costs are spread out among units.

Europe

 Europe collaboratively leads the cause for sourcing onshore and offshore winds through the European Wind Energy Association (EWEA). With over 700 members from almost 60 countries, the EWEA

---

[130] "Wind energy's frequently asked questions",
http://www.ewea.org/wind-energy-basics/faq/, Accessed 2/18/2015
[131] "Wind Energy Facts",
http://www.ewea.org/wind-energy-basics/facts/, Accessed 2/19/2015

is an international voice for promoting wind energy.[131] As of 2012, the European Union (EU) has established 106,000 MW of wind power to provide 7% of its overall power needs, which is equivalent to the electricity needs of Sweden, Ireland, Slovenia, and Slovakia combined.[131] In 2012, the EU invested close to 20 billion dollars in EU wind farms, and they show no signs of slowing down in investing in renewables like wind.[131] Currently, the EU's growth in wind has been motivated by the 2020 goal to hit three 20% targets: reducing EU greenhouse gases by 20% from 1990 levels; improving EU energy efficiencies by 20%; increasing EU's stake of power production from renewable resources to 20%.[132] Concerning wind, the EWEA estimates that wind power in Europe will double to 203,000 MW—20% of that figure coming from offshore wind.[131] In fact, more than one quarter of Denmark's electricity generation came from wind power in 2011.[131] Europe has invested greatly in wind, as well as other alternative powers, and for the most part, it seems like it has paid off by contributing billions of dollars to the EU's economy. Altogether, the fruitful areas for harvesting wind and the motivated and collaborative push for green energies make Europe the largest participant in wind power to date.

China

China's explosive growth has rocketed itself to the top of the charts in energy consumption, almost taking the crown from the United States. [133] China primarily uses coal and some oil for much of its energy needs.[133] Like the EU, China has challenged itself to increase its use of renewable energy sources, albeit at a more modest 15% by 2020.[133] In particular, China has boosted its hydroelectric infrastructure up to the point where it accounts for almost a quarter of its electricity capacity, the most of any country by megawatts.[133] The third largest provider of electricity in China is wind, making China the second-largest wind producer globally, again just slightly behind the United States.[133] According to the Global Wind Energy Council, in 2014, China actually had a higher overall capacity in wind power than any other country.[134] Overall, China's current installed capacity is 115,000 MW, and it plans

---

[132] "The 2020 climate and energy package",
http://ec.europa.eu/clima/policies/package/index_en.htm, Accessed 2/19/2015
[133] "China", http://www.eia.gov/countries/cab.cfm?fips=CH, Accessed 2/19/2015
[134] "Global Statistics" ,
http://www.gwec.net/global-figures/graphs/, Accessed 2/19/2015

to reach a capacity of 200,000 MW by 2020.[135] For instance, in the western province of Xinjiang near snowcapped mountains, giant wind farms are taking shape with manufacturers able to churn out two turbines per day due to large government subsidies.[135] Indeed, everything in China is on a massive scale, and their domestic wind resource is no exception.

The United States

The varied topography of the United States is apt for harnessing wind, with the Great Plains states having the highest concentration of wind turbines.[136] The Great Plains states from North Dakota and Montana to Texas receive on average higher wind speeds at turbine height than any other states.[136] As previously mentioned, the United States has long held the spot for consuming the most energy. Therefore, it only makes sense that they play a large role in shaping the world's energy economy. Currently, the United States has a fairly diverse portfolio of different energy resources for generating electricity: coal at 39%, natural gas at 27%, nuclear at 19%, and hydropower at 7%.[137] Among the newer renewables like biomass, solar, and geothermal, wind produced the most power at 4.13% in 2013.[137]

In 2004, the US wind sector produced only 14,000 MW, which is less than what is extracted from landfill gas and biogenic municipal solid waste today.[138] Nowadays, wind provides US homes tenfold what it did in 2004.[138] Similar to Europe and China, the United States has set itself a goal and is currently on track to produce 20% of its electricity from wind.[118]

There are several major influences on wind power in the United States that has allowed this enormous growth. Foremost, public outlook on energy and its effects has overwhelmingly broadened due to energy price booms and drops, climate change advocators and opponents, and

---

[135] "China on world's 'biggest push' for wind power"
http://www.bbc.com/news/science-environment-25623400, Accessed 2/21/2015
[136] "Wind generating capacity is distributed unevenly across the United States",
http://www.eia.gov/todayinenergy/detail.cfm?id=2470, Accessed 2/21/2015
[137] "What is U.S. electricity generation by energy source?",
http://www.eia.gov/tools/faqs/faq.cfm?id=427&t=3, Accessed 2/21/2015
[138] "Net Generation from Renewable Sources",
http://www.eia.gov/electricity/monthly/epm_table_grapher.cfm?t=epmt_1_01_a,
Accessed 2/21/2015

headline-events that shake the energy world. At the same time, many important decisions are being made in Washington D.C., which have lasting effects on how society deals with energy. For example, the Federal Production Tax has been the primary federal incentive for wind power.[118] As a result, since 2008 more than $100 billion in private investments have funneled into the United States for wind energy.[118] In addition, technological advancements have cut the cost of wind power by 58% over the past five years.[118]

Among individual states, Texas has shown the most development in wind. Specifically, wind energy supplied more than 10% of Texas's electricity in 2014.[139] Moreover, Texas has enough turbines to add more than 7,500 MW of capacity to the surrounding grid.[118] Oregon ranks sixth for installed wind capacity at 3,153 MW, and it shows moderate but steady growth.[140] Much of Oregon's wind farms are densely concentrated in Eastern Oregon, near the Columbia River Gorge. Overall, wind power generation has brought in over $6 billion in investments to Oregon.[140]

Society and Wind

In 2014, the international wind market grew by 44%, surpassing 50,000 MW in growth.[134] As evidenced by the large influx in investments into wind energy across the board, the global society almost unanimously supports augmenting the wind industry. Wind power has its fair share of both advantages and disadvantages that ultimately dictate the public's opinion.

The bottom line is that wind energy is a very green energy source. It does not produce any polluting emissions into the air, unlike power generating plants that run on the combustion of fossil fuels. Nor does it consume water during operation, so US turbines would theoretically reduce the total water use of the US electric sector by 8% from 2007 through 2030.[141] The actual resource of wind is free and ultimately sustainable literally as long as the sun shines. In addition,

---

[139] "Wind generates more than 10% of Texas electricity in 2014", http://www.eia.gov/todayinenergy/detail.cfm?id=20051, Accessed 2/21/2015
[140] "Oregon Wind Energy", http://awea.files.cms-plus.com/FileDownloads/pdfs/Oregon.pdf , Accessed 2/21/2015
[141] "Advantages And Challenges Of Wind Energy", http://energy.gov/eere/wind/advantages-and-challenges-wind-energy, Accessed 2/21/2015

wind is a domestic source that countries can produce to sustain themselves while reducing their dependence on foreign fuels. Advocates for battling climate change believe progressing wind energy is a vital component moving forward. Wind is indeed effective in displacing carbon emissions. For example, the US wind power generation avoids the equivalent amount of carbon pollution of 28 million cars.[118] Wind turbines can also be built on the millions of acres of US farmland, while taking up little footprint and financially supplementing landowners. Finally, wind power is competitively priced compared to other renewable energy technologies, costing between four and six cents per kilowatt-hour, based on the project and its funding.[141] These advantages explain why wind energy is the fastest-growing energy source worldwide.[141]

Nevertheless, wind has a handful of hurdles that hinder it from being the panacea to things like climate change and decline of fossil fuels. Chiefly, wind power costs must still improve to compete with conventional power generation sources such as natural gas. Second, ideal wind farm locations often are remote and far from cities where power is most needed. Even in the countryside, the footprint of a wind turbine may be better financially used for another purpose.[141] Further, the wind does not always blow, so power supplies will have to be backed up by another source. On the other hand, severe storms can shut down wind farms and disable power for the masses.

There have been various studies done on the environmental impacts of wind turbines.[142] In copious headlines, wind turbines have been often blamed for killing off birds. Although the number of birds killed by turbines is far less than those killed by cars and tall buildings, various organizations have spent years investigating how to minimize the damages in a cost-effective manner.[142] The result was better siting of wind turbines to avoid major bird migration. Even abroad, nature conservation groups like Birdlife and Greenpeace still support wind energy, because it helps to mitigate climate change.[130] Other common concerns are the noise and the aesthetics of turbines. The noise of turbines has been reduced since early models, and the noise levels generally comply with recommendations from World Health

---

[142] "Environmental Impacts And Siting Of Wind Projects", http://energy.gov/eere/wind/environmental-impacts-and-siting-wind-projects, Accessed 2/22/2015

Organization for living areas.[130] There have been some cases of wind farms obscuring scenic areas or disrupting historical sites, so governments have issued limitations on where turbines can be built.

Technology Frontier

New technology brought many of the advantages wind power has to offer, and it can also help solve many of the challenges wind power has to face. In the United States, three important areas relating to wind that could see improvement are the grid, storage, and turbine. The US grid must expand its reach to receive high supply from remote wind farms. Furthermore, all 200,000 miles of high-voltage transmission lines that make up the US grid can be improved by replacing them with more efficient thicker wires.[143] In order to integrate a variable power source like wind, the grid has to become smarter.

Luckily, the US grid is already capable at balancing variables. The US grid is separated into three relatively isolated interconnections: the Western Interconnection, the Eastern Interconnection, and the Electricity Reliability Council of Texas Interconnection.[143] The Western and Eastern Interconnections are broken further into balancing authorities.[143] Each balancing authority has a team of grid operators working around the clock to balance a sort of electric checkbook. That is, they ensure that the amount of power generated by various sources matches the demand. Large grids are actually more efficient and are fitter for introducing wind power, because they can draw from a larger pool of suppliers.[143] For instance, the PJM balancing authority serves 60 million customers across thirteen very populous Mid-Atlantic and Great Lake states.[143] Hence, when everyone cranks on their space heaters on a chilly night, the consumer demand increases, and the grid operators have to react accordingly by calling power plants to increase supply. Therefore, the power grid can handle variable demand, and variable supply is not too far off considering that numerical weather prediction is improving. Incidentally, researchers can also simulate how wind turbines affect weather systems. One Stanford study even explored how off shore turbines could dampen hurricanes.[144] In short, even though the

---

[143] "Transmission 101",
http://www.awea.org/Issues/Content.aspx?ItemNumber=866, Accessed 2/23/2015
[144] "Offshore wind farms could tame hurricanes before they reach land",
http://news.stanford.edu/news/2014/february/hurricane-winds-turbine-022614.html, Accessed 2/24/2015

US grid can deal with variability and can improve predicating said variability; it will still have to lay a large number of transmission lines to remote areas.

Alternatively, the Department of Energy is also exploring distributed wind energy systems. These systems are made up of smaller ranged turbines from a single 5-killowatt turbine powering a house to a couple turbines outputting less than 20 MW that can be installed within a city's grid.[145] They are connected on the customer side of an electric meter, so individuals can invest in a turbine, use the generated power, and sell excess power to the grid.[145]

In any case, the grid would benefit from having a way to store renewable energies. Enhancing the current energy storage system is crucial in order to make wind power more independent. There are several branches of energy storage, including but not limited to, compressed air, pumped hydroelectricity, and magnetically levitating flywheels. At the top of the storage food chain are batteries, which have undergone iterative optimization for different uses. Numerous wind farms already utilize battery arrays. Simply put, batteries are dense, efficient, and responsive. Battery storage, however, is many times more expensive than pumping water into a reservoir, one of the cheapest storage options.[146] A study on wind and storage has demonstrate that when modeling the Colorado power grid with 10% of its power from wind, a 324-MW hydroelectric pumped storage plant would save $2.5 million a year.[146] Still, it would take over a hundred years to pay back the storage plant's costs and to financially benefit from it.[146] All in all, currently at wind power's small share, storage is not necessary to assimilate wind with other energy sources.

On the turbine side, the main path is growing the turbine's physical size and developing its control systems. As mentioned before, turbines are expected to double in size to 250 meters in diameter.[123] To put this in perspective, a single turbine this size could have a rating of 20 MW as it sweeps an area close to the size of the base of the Great Pyramid of Giza.[123] The generic design of the turbine is likely here to stay, but there are several system design improvements in the works.

---

[145] "How Distributed Wind Works",
http://energy.gov/eere/wind/how-distributed-wind-works, Accessed 2/24/2015
[146] "Wind energy and storage",
http://www.awea.org/Issues/Content.aspx?ItemNumber=5452, Accessed 2/24/2015

First, turbines are being developed for assorted weather and climate conditions, like icy climates, tropical storms, and even low-wind conditions.[123] In general, system engineers are fine-tuning the costs and performance of wind plants, including achieving the lowest LCOE.[123] Research and development is already underway to upscale turbines from 10 MW to 20 MW.[123] Furthermore, pushing the limits of turbines necessitates using advance modeling with higher accuracy as well as establishing testing facilities for novel designs.[123]

One interesting wind turbine prototype to come out of the works is a floating wind turbine. The wind turbine sits atop a triangular platform that can be moored with cabling. The selling point of floating wind turbines is that they can be stationed in deep waters where otherwise the anchoring of conventional offshore turbines would be costly or unfeasible.[147] Since the turbines are farther from land, they are less of a visual nuisance for onlookers and less intrusive in fishing lanes. There are a handful of full-scale prototypes in the coasts of Portugal, Norway, and Oregon.[147]

Flying wind turbines are yet another unusual group of turbines. The thought is that at higher altitudes the wind is much more consistent and stronger. The Buoyant Airborne Turbine, or BAT for short, is a cylindrical blimp-like structure that houses a rotor and blades inside.[148] It is unmanned at about 1,000 feet above the ground, but it can be remotely controlled and adjusted with tethers.[148] The BAT can potentially provide power to places where electricity is scarce or expensive; because it is portable enough to be shipped in two shipping containers and can operate with minimal worker oversight.[148] The BAT is not the only turbine to take for the skies.[148] The PowerShip is actually a blimp with turbines on either side.[148] The Makani is a tethered airplane that flies continuously in circles.[148] The Enerkite is truly just a large kite that spools a turbine.[148]

Wind's Future

All of the aforementioned prototypes reflect the continuing innovations in wind power. Windmills have evolved immensely from

---

[147] "Windfloat",
http://www.principlepowerinc.com/products/windfloat.html, Accessed 2/25/2015
[148] "Meet the BAT, an airborne wind turbine",
http://www.cnn.com/2014/05/12/tech/innovation/big-idea-airborne-wind-turbines/,
Accessed 2/25/2015

pumping water for livestock and grinding grain. Nowadays, turbines still conduct the simple task of harnessing wind, but they are packed with technology and years of tinkering and optimizing. Turbines have moved to the ocean and to the skies. As an energy source, wind is not only environmentally friendly but also economically sound. As a testament to wind, the world's major consumers and producers, like Europe, China, and the United States, have all invested significantly in developing wind. Moreover, these investments have returned a stable energy source and created broader job opportunities. In order to further develop wind into a major power producer for the United States, work needs to be done to establish more wind farms without harming ecological and social facets. In every instance, from nation to nation, federal incentives were the driving force behind expanding wind, and they can help dampen the initial costs of wind. Wind power's future is in the hands of producers and consumers alike to dig a little deeper into their pockets and invest in wind energy matters.

Problems

1. Air moves due to a difference in what?
   Answer: Air pressure
2. If wind speeds increased by 20% at night, what is percentage increase of the power output using the idealized equation 9.3.
   Answer: $P \propto v^3 \rightarrow 120\%^3 = 172.8\%$
   72.8% increase or 172.8% of the original power.
3. How many 80-meter (diameter) turbines would it take to provide 3.7 MW of wind power in Corvallis, OR, with winds speeds on average at 3.5 meters per second assuming the Betz limit and 1.2 kg per meter cubed for air density?
   Answer: Using equation 9.3, solve for the individual power output of turbine with those parameters. Then, solve for the number of turbines it would take to output 3.7 MW.

$$P_i = \frac{1}{2}\left[1.2\frac{kg}{m^3}\left((80 \text{ m})^2\frac{\pi}{4}\right)\left(3.5\frac{m}{s}\right)^3 * .59\right] = 76{,}292 \text{ W}$$

$$\#\text{Turbines} = \frac{3{,}700{,}000 \text{ W}}{76{,}292 \text{ W}} \sim 49 \text{ turbines}$$

4. The National Renewable Energy Laboratory predicts in 2025 onshore wind power will increase to 42%, overnight capital costs will drop to 1,600 \$/kW, and fix O&M costs will drop to 20 \$/kW-yr. Recalculate the simple LCOE for these parameters. (*Assuming CRF of 10%*).

$$\text{sLCOE} = \frac{1{,}600\frac{\$}{kW} * \frac{0.10}{yr} + 20\frac{\$}{kW * yr}}{8760\frac{hr}{yr} * (.42)} + \text{O\&M}_v + \text{FC} * \text{HR}$$

$$= 0.0489\frac{\$}{kWh} \rightarrow 4.89\frac{cents}{kWh}$$

5. Define capacity factor:
   Answer: "Capacity factor is a measure of how often an electric generator runs for a specific period of time. It indicates how much electricity a generator actually produces relative to the maximum it could produce at continuous full power operation during the same period." From EIA

Chapter 9. Geothermal Energy
By Brian Bemis

Geothermal energy is a form of renewable energy with the potential to produce energy on a large scale with many benefits. Therefore, it is important to be aware of how it works; what types of geothermal energy there are; and how it interacts with the economy, the environment, and society. It is also advantageous to be aware of what improvements the field can make. Geothermal energy is based on the idea of obtaining heat from the earth. This is accomplished by taking water stored in underground reservoirs that is heated by the earth's core. When extracted from the earth, this hot water becomes steam that can be used to turn a turbine.

Thermal Cycle

The thermal cycle for geothermal energy is similar to other energy sources, but it does have unique aspects to it. The process begins by drawing hot water from the subterranean reservoir. The steam obtained by this process is used to move a turbine to generate electricity. Then the steam is condensed and returned to the earth. Geothermal energy is unique, because the earth acts as both a heat source and a cold sink for energy production. This creates a potentially completely renewable resource.

Dry Steam

The simplest of the three types of geothermal power is called dry steam. Dry steam was the first source of geothermal power generation. The first use of geothermal power was at Lardarello in Italy in 1904.[149]

The dry steam method is utilized at locations where hot steam exits the earth. Places such as geysers are good examples of these locations. Dry steam geothermal plants use the hot steam produced by the earth by directing it to directly turn a turbine to generate electricity.[150]

---

[149] "Types of Geothermal Power Plants." Energy Almanac. Accessed March 1, 2015,
http://energyalmanac.ca.gov/renewables/geothermal/types.html.
[150] "Types of Geothermal Power Plants." Energy Almanac. Accessed March 1, 2015,
http://energyalmanac.ca.gov/renewables/geothermal/types.html.

Flash Steam

The next type of geothermal power plant is the flash steam system. Today, it is the most commonly used system for creating geothermal powered electricity. This is due to the system's balance of technological freedom and simplicity. This system builds off of the design of the previous system, the dry steam system.

The flash steam system works similarly to the dry steam system. This process uses geothermal heat to boil the water. AN injection well is drilled and water is pumped into the earth. The water flows into fractures in the hot rock. Then a production well is drilled and the now hot water is pushed out of the earth and out of the production well. This push is caused by the pressure of the water entering the injection well in conjunction with the same high subterranean pressure that causes geysers to erupt. The hot water that is retrieved from the earth is called brine. As the brine rises to a lower pressure environment it begins to evaporate. The part of the brine that is now steam is harvested to turn a turbine and generate electricity. This steam is separated from the still liquid brine by a flash system. Later, the liquid brine and condensed steam are returned to the injection well and pumped back into the earth reservoir to complete the cycle.

Another version of this type of geothermal power plant is the double flash steam cycle. This system follows the same process as its namesake until the liquid brine is separated from the steam. This liquid brine is then sent to more flash systems in a succession of lower pressures until a large percentage of the brine has been converted into steam. From there the steam and little remaining liquid brine follow the same processes of spinning a turbine and returning to the earth through an injection well.[151]

Binary Cycle

The third geothermal system is the binary cycle. The National Renewable Energy Laboratory reports this cycle operates with water at lower temperatures of about 225 °F – 360 °F and is the more experimental type of geothermal power.

---

[151]Valdimarsson, Páll. "GEOTHERMAL POWER PLANT CYCLES AND MAIN COMPONENTS." 2011. Accessed March 1, 2015. http://www.os.is/gogn/unu-gtp-sc/UNU-GTP-SC-12-35.pdf.

The binary cycle system can be thought of as an evolution of the flash system. This process uses the hot subterranean environment to heat brine pumped into the earth through an injection well and retrieved through a production well. But once the brine enters the aboveground facilities the similarities between the cycles end.

The binary cycle has an additional second step, hence the term "binary" in the name. In the flash steam system, once the brine leaves the high-pressure of the earth it evaporates due to its high temperature. However, the binary cycle is designed to take advantage of brine that cannot reach a hot enough temperature to boil. This is accomplished through the use of a heat exchanger.

The heat exchanger transfers the heat from one substance to another.[152] This machine transfers the heat of the brine to another substance called a working fluid. Working fluids are chemically designed to have low boiling points so that the brine's heat is enough to evaporate the fluid into steam that is used to turn a turbine and generate electricity. It is important to note that the working fluid is contained separately from the brine and is never allowed to leave its containment. This allows the working fluid to be created using different chemical compositions. Most of these working fluids consist of hydrocarbons and other chemicals. Also, because the working fluid is contained, the binary cycle system restricts gas emissions. The only air emissions are water vapor from the heated brine.

There are three methods of obtaining brine that can be utilized in a binary cycle geothermal power plant. These methods are: Enhanced Geothermal Systems (EGS), Low-Temperature Resources, and Co-Produced Resources.

The EGS is designed to improve the geothermal reservoir where the brine is heated by increasing the surface area of the subterranean rock. This is accomplished by using the brine to enlarge the fissures through force or by dissolving the rock. This increases the efficiency of the geothermal power plant by making the rock more permeable so more brine can flow through and be heated. This process is designed to

---

[152] *Random House Webster's College Dictionary*, 2nd Rev. and Updated Random House ed. New York, New York: Random House, 2001. 608.

artificially enhance a potentially viable geothermal area so that it can be used to produce enough brine to run a geothermal power plant.[153]

There are three types of EGS: in field, near field, and green-field. In-field types are designed to deal with small areas that are already part of an otherwise usable geothermal location. The near-type is designed to operate near a usable location's outer margins. Green-field types deal with areas where geothermal resources have yet to be utilized.[154]

Low-Temperature resources are geothermal sites that heat brine to temperatures of less than 300 °F. These resources are important, because they are far more common and easier to harvest than other geothermal resources. This resource is also commonly used in a direct-use system, because such systems do not require high temperatures for optimum efficiency.[155]

Co-Produced Resources are different from the other sources of geothermal brine. These resources originate as a byproduct from the oil and gas well facilities. Instead of being disposed of, this resource can be used in a binary cycle to minimize emissions and generate power. This process can be utilized to make oil and gas wells more eco-friendly.[156]

Advantages and Disadvantages of Each Plant Type

There are advantages and disadvantages to using each of the three geothermal power plant systems. The dry steam, flash steam, and binary cycle systems should be analyzed in turn with respect to the other systems.

The dry steam system was the first to be designed and is the most straightforward of the three systems. However, the dry steam system is severely limited to specific locations. It can only be utilized in areas with the proper type of steam vents and there are only two known locations in the United States that are sufficient to supply a geothermal power plant purely from this method.

---

[153] "How an Enhanced Geothermal System Works." Energy.gov. Accessed March 1, 2015, http://energy.gov/eere/geothermal/how-enhanced-geothermal-system-works.
[154] MATEK, BENJAMIN. "The Manageable Risks of Conventional Hydrothermal Geothermal Power Systems." 2014. Accessed March 1, 2015. http://geo-energy.org/reports/Geothermal Risks_Publication_2_4_2014.pdf.
[155] "Low-Temperature and Coproduced." Energy.gov. Accessed March 1, 2015. http://energy.gov/eere/geothermal/low-temperature-and-coproduced.
[156] "Low-Temperature and Coproduced." Energy.gov. Accessed March 1, 2015. http://energy.gov/eere/geothermal/low-temperature-and-coproduced.

With the technological innovations in drilling methods, the flash steam system was developed. This system is more complicated than the dry steam system, as it requires the drilling of wells and running a pumping system. It is advantageous because it recycles the brine and creates a renewable resource. However, it is also restricted by location, but it has more possibilities than the dry steam systems. This system does come with environmental concerns that are stronger than the dry steam system had. The main cause of this concern is the possibility of the brine transferring foreign substances into the wrong environment.

The binary cycle system is the current system in development. This system adds another layer of complexity to the flash steam system by adding a heat exchanger. The pay-off is that it can function with locations that can only heat the brine to 300 °F. This fact and the use of Enhanced Geothermal Systems (EGS) allow even more locations to become geothermal power plants. This means that the binary cycle system is less restricted by location than the dry steam and flash steam systems. This form of geothermal power is currently undergoing the most experimental growth as it has multiple sources of power being tested, notably the low-temperature and co-produced resources. However, it also has the same brine potential for environmental problems that the flash steam system has. The EGS version of geothermal plants raise concerns about potential seismic issues as the process of opening underground fissures causes seismic activity. The working fluids that are used also pose an issue if not dealt with properly as they should not be exposed to the environment.

Direct-Use

Another type of geothermal power is called direct-use. Unlike the majority of this chapter, the direct-use system is not involved with electricity generation in a geothermal power plant. Direct-use may not have a large-scale power generation capacity to add to the electric grid, but it is important to know what it is, what its benefits are, and what its short-comings are.

The direct-use system involves using geothermal fluids to provide direct heat. It possesses three components: the production facility, mechanical system, and disposal system. It is used to replace traditional heating systems in groups of individual buildings by supplying the hot water to be pumped throughout the buildings. Direct-use also has applications in agriculture and commercial circumstances.

A simple example of this is that greenhouses can save about 80% of their fuel costs by using direct-use geothermal energy. [157] Food dehydration is the most common use of geothermal energy in the industrial and commercial world. The cool geothermal fluid can either be recycled back into the system or it can be used as water such as for plant irrigation or washing purposes. In many cases the direct-use resource is combined with a binary cycle so that different fluids can be sent to consumers, such as clean water for home use and swimming pools.

There are many benefits to the direct-use system. The first is that it can be used in many different locations compared to geothermal power plants. This is due to the fact it can use geothermal fluids at temperatures of 68 °F to 302 °F (20 °C to 150 °C). Compared to a power plant, the direct-use system can function in locations with smaller geothermal resources. This is shown by an identified 1,277 geothermal sites within 5 miles of 373 cities in 8 states. [158]

Direct-use geothermal energy does have its limits, though. Most of the sites that can provide geothermal energy are in the western United States. The sites are not large enough to generate a meaningful amount of electricity and are therefore restricted to the aforementioned uses. This means that although it is beneficial to individualized areas it will most likely never be a large-scale energy resource.[159]

Economic Costs

It is easy to see the advantages of geothermal power, but such advantages do not come free. There are two major monetary costs associated with geothermal power development and maintenance costs. It is also important to consider how much time will be needed to make up for those costs. Therefore the cost of geothermal energy is important to consider.

Development costs are the highest percentage of the cost of geothermal power. The costs include: surveying areas to find suitable locations, drilling multiple wells, constructing the pipelines, equipment

---

[157] "Direct Use of Geothermal Energy." Energy.gov. Accessed March 1, 2015, http://energy.gov/eere/geothermal/direct-use-geothermal-energy.
[158] "Direct Use of Geothermal Energy." Energy.gov. Accessed March 1, 2015, http://energy.gov/eere/geothermal/direct-use-geothermal-energy.
[159] "Direct Use of Geothermal Energy." Energy.gov. Accessed March 1, 2015, http://energy.gov/eere/geothermal/direct-use-geothermal-energy.

to analyze the subterranean reservoir's status and finally constructing the plant. The initial cost for the field and power plant is around $2,500 per installed kW in the U.S., probably $3,000 to $5,000/kWe for a small (<1MWe) power plant. [160] Maintenance costs are not as high as the initial costs. Generally, maintenance costs range from $0.01 to $0.03 per kWh. [161]

Payback period is the period of time required to recoup a capital investment. [162] For a geothermal power plant, it varies based on the source's flowrate and temperature. The desired payback period when establishing a plant is less than ten years. The flow rate is more impactful than temperature on how long the payback period of a geothermal power plant will be. [163]

The payback period and other economical measurements will improve as the amount of heat the geothermal power plant produces increases. One method to denote a plant's ability is to calculate the heat delivery it possesses. Provided by to the European Geothermal Energy Council, the equation to calculate heat delivery is:

$$P = \rho \; C_p \; V \; (t_{in} - t_{out})$$

P is the heat delivery [J/h], ρ is the volumetric mass [kg/m³], C_p is the mass heat capacity [J/(kg*K)], V is the flow volume [m³/h], and t_{in} and t_{out} are the fluid temperatures [K]. In many cases geothermal power plants are required to measure the fluid temperatures and flowrate [164]

For example, consider a geothermal plant that receives brine from a production well with a flow volume of 45 m³/s and a volumetric mass of 1 g/cm³ at 300 °C and returns the brine to an injection well at 40

[160] "Geothermal FAQs." Energy.gov. Accessed March 1, 2015, http://energy.gov/eere/geothermal/geothermal-faqs.
[161] "Geothermal FAQs." Energy.gov. Accessed March 1, 2015, http://energy.gov/eere/geothermal/geothermal-faqs.
[162] Random House Webster's College Dictionary, 2nd Rev. and Updated Random House ed. New York, New York: Random House, 2001. 973.
[163] Fitzgerald, Crissie D. "An Economic Evaluation of Binary Cycle Geothermal Electricity Production," *Air Force Institute of Technology Theses*, 2003, 53.
[164] "Key Issues for Renewable Heat in Europe." *EUROPEAN GEOTHERMAL ENERGY COUNCIL*: 9. Accessed March 1, 2015. http://www.erec.org/fileadmin/erec_docs/Projcet_Documents/K4_RES-H/D8_EGEC.pdf.

°C. The brine has a heat capacity of 5 kJ/(kg*K). Find the plant's heat delivery.

Solution:

Process:

1. Consider given values and desired values.
2. Convert values to their proper units for the formula.
3. If necessary adjust formula to solve for desired value.
4. Input the given values and solve.

Work:

1. Given: V= 45 m³/s, ρ =1 g/cm³, t_{in}=300 °C, and t_{out} = 40 °C, C_p =5 kJ/(kg*K). Want to find: The heat delivery in J/h.

2. Unit conversion:

   a. $V = \dfrac{45 \text{ m}^3}{s} \dfrac{3600 \text{ s}}{h} = 162{,}000 \text{ m}^3/h$;

   b. $\rho = \dfrac{1 \text{ g}}{cm^3} \dfrac{(100)^3 cm^3}{m^3} \dfrac{kg}{1{,}000g} = 1{,}000 \text{ kg/m}^3$;

   c. $t_{in} = 300\,^{\circ}C = (300 + 274.15)K = 574.15 \text{ K}$;

   d. $t_{out} = 40\,^{\circ}C = (40 + 274.15)K = 314.15 \text{ K}$;

   e. $Cp = \dfrac{5 \text{ kJ}}{(kg*K)} \dfrac{1{,}000J}{kJ} = 5{,}000 \text{ J/(kg} * K)$.

3. $P = \rho \, Cp \, V \, (t_{in} - t_{out}) =$

   $1{,}000 \,{}^{kg}\!/_{m^3} \; 5{,}000 \,{}^{J}\!/_{(kg \, * \, K)} \; 162{,}000 \,{}^{m^3}\!/_{h} \times (574.15 \text{ K} - 314.15 \text{ K}) =$

   $1{,}000 \,{}^{kg}\!/_{m^3} \; 5{,}000 \,{}^{kg \, \cdot \, m^2}\!/_{s^2 (kg \, \cdot \, K)} \; 162{,}000 \,{}^{m^3}\!/_{h} \; (260 \text{ K}) =$

   $2.1E13 \dfrac{kg \cdot m^2}{s^2 \cdot h} = 2.1E13 \text{ J/h} = 2.1E7 \text{ MJ/h}$

Therefore, the geothermal power plant's heat delivery is 21,000,000 MJ/h.

Economic Benefits

As with all types of renewable energy, the economy is a major driving force in deciding how much time, money, and research is invested in advancing the technology. There are several economic benefits to geothermal power. These include new markets, government income, and variability.

Geothermal power is a field that is still developing. This means there is opportunity for businesses to succeed, or fail. Due to government subsidies, the financial risks of entering the field are not as high. Nonprofit research facilities are already developing new technology to advance the field.

Governments can also benefit directly from geothermal energy. The Geothermal Energy Association states that about half of geothermal plants operate on public lands generating revenue for state, municipal, and federal governments.[165] Thus geothermal power plants can be an important source of revenue for governments.

Geothermal power plants can be a firm or flexible power source. This allows them to provide the grid with a versatile supply of energy. This is beneficial to the hectic balance of the grid's energy consumption and intake.

Environment

Geothermal power plants have an important relationship with the environment. One of the main selling points of geothermal energy is its benefit to the environment, therefore it one must seriously consider the environment when dealing with geothermal power. Geothermal energy has positive aspects, negative aspects, and fixable negative aspects.

There are many qualities about geothermal energy that are beneficial. One is that geothermal energy produces almost no carbon dioxide and other greenhouse gas emissions. Geothermal power plants require less land space than other types of energy plants. This is because the process that provides the heat source is below ground. Geothermal energy also benefits from its environmental aspects in that unlike the other renewable sources of energy: wind, wave, and solar; geothermal energy is capable of running all the time regardless of the weather.

On the other hand, geothermal power plants also pose potential threats to the environment. One potential threat is that as the brine moves through the earth it collects minerals that can be harmful to ecosystems if the brine leaks into waterways by seeping into underground rivers. According to The Encyclopedia of New Zealand, geothermal fluids contain elevated levels of arsenic, mercury, lithium, and boron. If these waste minerals are released into rivers or lakes these pollutants can damage aquatic life and make the water unsafe for

---

[165] Matek, Benjamin, and Karl Gawel. "The Economic Costs and Benefits of Geothermal Power." 2014, 7, http://geo-energy.org/reports/Economic%20Cost%20and%20Benfits_Publication_6_16.pdf

drinking or irrigation.[166] There is concern that dangerous seismic activity could occur while the brine is introduced to the subterranean reservoir. This is even more likely during the creation process of EGS. Another concern is a potential depletion of the earth's heat resource. This can occur if a geothermal plant is not properly managed and this could interfere with the surrounding environment. A visible concern is the potentially damaging effects geothermal power plants can have on geological features. This is mostly a concern with dry steam plants that affect natural geysers. A dangerous possibility can occur if too much pressure is released from the production wells. This can lead to the area of land around the geothermal plant sinking due to a lack of underground pressure. Geothermal energy does have a small potential to release carbon dioxide and other harmful gasses, because the working fluids in a binary cycle usually contain hydrocarbons. Some of these can escape into the atmosphere when the working fluid gas is used to turn the turbine above ground.

Fortunately, there are solutions to most of these problems. Unfortunately many of these problems rely on human care and analysis of potential sites in order to avoid issues. The selection of a site either far from aquifers or with impermeable rock insulating the reservoir will prevent the brine from carrying foreign minerals into the underground rivers. Selecting a sturdy site and using care when drilling will limit the potential seismic activity and subsidence. To prevent depletion of the earth's heat resource calculations can be made to determine how long each period of rest and use should be for a plant so that the site can regenerate lost heat. [167] In order to prevent damaging important geological features, geothermal sites are not built at such sites altogether. For example, Yellowstone a potential location for a productive dry steam geothermal plant, but it is protected and therefore no geothermal power plant will be built there. The solution to escaping gas state working fluids is to develop a better containment system for the working fluid cycle.

---

[166] Stewart, Carol. "Geothermal Energy - Effects on the Environment." *Te Ara - the Encyclopedia of New Zealand*, 2012, 5, Accessed March 1, 2015. http://www.teara.govt.nz/en/geothermal-energy/page-5.
[167] "Electrical Power Generation from Geothermal Sources." Battery and Energy Technologies. Accessed March 1, 2015.
http://www.mpoweruk.com/geothermal_energy.htm.

Management Groups

There are many groups that manage geothermal energy. These groups deal with the different aspects of geothermal energy. They can be divided into three categories the government, nonprofit, and business.

There are many governments that actively manage geothermal energy. The role of government is to accurately analyze geothermal energy. This allows important comparisons between different types of geothermal power and other types of power sources to be made. Governments also provide financial aid to encourage the advancement of renewable resource technologies such as tax incentives and loan guarantees.[168]

In the past, the main way the United States government has made contributions to geothermal energy is through the use of federal acts. The earliest act was the Geothermal Steam Act, which allows the government to lease public land for geothermal advancement. The next was the Geothermal Energy Research Development and Demonstration (RD&D) Act, which provides secure loans towards geothermal energy endeavors. Then the Energy Policy Act of 2005 was signed into law to change U.S. energy policy by providing tax incentives and loan guarantees to encourage the construction of geothermal facilities. The Advanced Geothermal Research and Development and American Recovery and Reinvestment Acts provide authorization and funding respectively for geothermal research.[169]

GeoPowering the West is an initiative by the US Department of Energy for identifying and developing geothermal technologies. It focuses on the more geothermal prevalent western states, but it works on a national level. This initiative provides resources to private businesses working with geothermal energy technology.[170]

The National Science Foundation (NSF) is an independent federal agency that supports advancing many forms of scientific research, including geothermal energy production. The primary way the NSF supports geothermal research is by funding research projects in

---

[168] "A History of Geothermal Energy in America." Energy.gov. Accessed March 1, 2015, http://energy.gov/eere/geothermal/history-geothermal-energy-america.
[169] "A History of Geothermal Energy in America." Energy.gov. Accessed March 1, 2015, http://energy.gov/eere/geothermal/history-geothermal-energy-america.
[170] "U.S. Department of Energy - GeoPowering the West (GPW)." Geothermal Energy Association, Accessed March 1, 2015. http://geo-energy.org/states_contacts.aspx.

different fields of science. This usually occurs at the college and university level.[171]

The U.S. Department of Energy (DOE) is the United States government's solution to address its responsibilities regarding energy production. As such the DOE provides extensive databases on all forms of energy production including geothermal energy. This group and its Office of Energy Efficiency and Renewable Energy and the Geothermal Technologies Office provide official data on numerous aspects of geothermal energy. The DOE also provides informational opportunities for people to learn more about geothermal energy.

There are different types of nonprofit groups that contribute to geothermal energy. One type is the private individual who uses direct geothermal energy to heat their buildings. Another type is the group that operates their own geothermal energy resource and researches it to improve geothermal energy as a whole.

The Geo-Heat Center in Klamath Falls Oregon is an example of one of these groups.[172] This group is affiliated with the Oregon Institute of Technology (OIT), also called Oregon Tech, but it records information on geothermal energy in many other locations. It specifically focuses on the geothermal at Oregon Tech.

The geothermal facilities at Oregon Tech consist of seven production wells and two geothermal power plants. Currently, only four of these wells are operating. Oregon Tech has used direct-use geothermal heating since 1964.[173] In 2010, it officially completed its first geothermal power plant with a maximum installed capacity of 280 kilowatts. [174] Later in December 2014, Oregon Tech finished constructing a second larger 1.75 MW geothermal power plant.[175]

---

[171] "NSF at a Glance." US NSF. Accessed March 1, 2015.
http://www.nsf.gov/about/glance.jsp.
[172] GEO-HEAT CENTER. February 6, 2012. Accessed March 2, 2015.
http://geoheat.oit.edu/.
[173] GEO-HEAT CENTER. February 6, 2012. Accessed March 20, 2015.
http://geoheat.oit.edu/ service.htm.
[174] Maupin, Kristina, and John Lund. "OREGON'S FIRST GEOTHERMAL COMBINED HEAT AND POWER PLANT DEDICATION." *GEO-HEAT CENTER Bulletin*, 2010, 1. Accessed March 20, 2015. http://geoheat.oit.edu/bulletin/bull29-1/art5.pdf.
[175] "Clean Energy Campus." Oregon Tech. Accessed March 20, 2015.
http://www.oit.edu/sustainability/clean-energy.

The geothermal power plant utilizes a combination of a binary cycle and a direct-use system. This plant uses four production wells to retrieve brine from the earth and then filters the brine so that its acidity does not interfere with the plant's technology. The brine is sent to a boiler where in is used to flash a working fluid called a refrigerant into steam. The refrigerant steam is used to turn a turbine to generate electricity. The refrigerant is then sent to a cooling tower before returning to the boiler. From the boiler the brine is sent to the university to be used in a direct-use system. Finally, the cooled brine is returned to the earth through a single reinjection well.

The cooler the refrigerant is when it returns from the cooling tower, the more efficient the geothermal power plant's energy production. Therefore, the geothermal power plant can produce more energy when it is cold outside. This quality is utilized by Oregon Tech to complement their solar array field, which understandable functions better during the heat of the day and not at all during the cool nights.

Oregon Tech plans to continue improving their geothermal system. One future improvement is to shut down all the production wells save the newest and biggest well. This is due to a concept called parasitic load. Parasitic load considers the fact that each of the production wells cost energy to function. Therefore, the future goal is to be able to run the geothermal facilities off of one production well thereby minimizing production costs. The concept of parasitic load is responsible for the pumps at the OIT facility running at only 60% capacity. The concept being that if they run any higher their energy cost will neutralize the resulting energy gain from the geothermal plant and the pumps will wear out faster.

The Geothermal Resources Council (GRC) is another nonprofit group dedicated to advancing geothermal energy. According to its website, the "... GRC actively seeks to expand its role as a primary professional educational association for the international geothermal community".[176]

The final group that manages geothermal energy is the private sector. Geothermal power plants are typically run by small businesses or a collection of small businesses. The Geothermal Energy Association is one of the main groups of businesses that manage geothermal power in

---

[176] "About the GRC - Geothermal Resources Council." Geothermal Resources Council. Accessed March 1, 2015. http://www.geothermal.org/about.html.

the United States. According to the GEA's website, "...the GEA advocates for public policies that will promote the development and utilization of geothermal Resources, provides a forum for the industry to discuss issues and problems, encourages research and development to improve geothermal technologies, presents industry views to governmental organizations, provides assistance for the export of geothermal goods and services, compiles statistical data about the geothermal industry, and conducts education and outreach projects".[177]

Location, Location, Location

One of the most important factors for building a geothermal power plant is the location where it can be built. The location must meet requirements that are different for each type of geothermal plant. There are locations throughout the world that are being used to generate geothermal energy. There are also locations that could produce geothermal energy that are not currently being utilized.

It is important to begin with an understanding of what is required for a location to be suitable for geothermal power. According to the U.S. Department of Energy's Office of Energy Efficiency and Renewable Energy a location should have "[h]ot geothermal fluid with low mineral and gas content, shallow aquifers for producing and reinjecting the fluid, location on private land to simplify permitting, proximity to existing transmission lines or load, and availability of make-up water for evaporative cooling. Geothermal fluid temperature should be at least 300 °F, although plants are operating on fluid temperatures as low as 210 °F".[178]

There are calculations that can be made to determine the requirements and possibilities of a location. These calculations use the formulas:

$$\text{Well Depth} = \frac{\text{Resource Temperature}(\,^\circ\text{C}) - \text{Ground Temperature}(^\circ\text{C})}{\text{Geothermal Gradient}(^\circ\text{C/km})} \times \frac{1,000 \text{ m}}{\text{km}}$$

where

Resource Temperature $(^\circ\text{C}) =$

---

[177] "About GEA." Geothermal Energy Association. Accessed March 1, 2015. http://geo-energy.org/aboutGEA.aspx.
[178] "Geothermal FAQs." Energy.gov. Accessed March 1, 2015, http://energy.gov/eere/geothermal/geothermal-faqs.

$$\text{Well Depth(m)} \times \frac{\text{km}}{1{,}000 \text{ m}} \times \text{Geothermal Gradient}(°C/\text{km}) +$$
$$\text{Ground Temperature } (°C).^{179}$$

These formulas allow estimations to be made on whether a location can maintain a geothermal power plant.

For example, consider a location with a geothermal gradient of 0.025 °C/m and a ground temperature of 572 °F. How deep would a well need to be to achieve a resource temperature of 662 °F.

Solution:

Process:

1. Consider given values and desired values.
2. Convert values to their proper units for the formula.
3. If necessary adjust formula to solve for desired value.
4. Input the given values and solve.

Work:

1. Given: Geothermal gradient = 0.025 °C/m , Ground temperature = 572 °F, and Resource temperature = 662 °F. Want to find: the well depth in meters.

2. Unit conversion:
   a. Geothermal gradient $= 0.025 \frac{°C}{\cancel{m}} \times \frac{1{,}000 \cancel{m}}{\text{km}} = 25 \, °C/\text{km}$
   b. Ground temperature $= (572 - 32) \times 5/9 = 300 \, °C$
   c. Resource temperature $= (662 - 32) \times 5/9 = 350 \, °C$

3. Well Depth $= \frac{350°C - 300°C}{25°C/\text{km}} \times \frac{1{,}000 \text{ m}}{\text{km}} = \frac{-50{,}000 \cancel{°C} \times \cancel{\text{km}} \times \text{m}}{25 \cancel{°C} \times \cancel{\text{km}}} = 2{,}000 \text{ m}$

Therefore, in order to obtain a resource temperature of 662 °F (350 °C) the well would need to be 2,000 m deep.

Next, it is important to note areas where geothermal power is in use. Major geothermal power plants are primarily in use in the Western United States, Iceland, Northern New Zealand, and Indonesia. The Geysers in Northern California are dry steam type geothermal power plants that are the most developed geothermal energy resource in the United States. The Geothermal Education Office states that Reykjavik ("Bay of Steam"), the capital of Iceland, with more than 145,000 people, pipes hot water to every house at a cost less than cold water. Wairakei in

---

[179] Fitzgerald, Crissie D. "An Economic Evaluation of Binary Cycle Geothermal Electricity Production", Air Force Institute of Technology Theses, 2003, 13, 33.

New Zealand was the first large hot-water field ever developed. These areas of the world are seeing the majority of focus into geothermal power; therefore it is likely that advancements will come from sources in these countries.[180]

For the sake of expanding geothermal power it is necessary to consider different viable locations to build a geothermal power plant. A 2009 map the National Renewable Energy Laboratory shows that the area around the Oregon-Idaho-Nevada border is highly favorable for creating new geothermal power plants. This study was specifically designed to implement the new EGS type of geothermal energy discussed earlier in the chapter. [181]

In summary, geothermal power plants are heavily dependent on location. Therefore it is necessary to consider the requirements of a location that can produce geothermal power, what locations are already harboring geothermal power plants, and what locations can have power plants constructed in the future.

Recommendations for Improvement

Geothermal energy was a developing resource, but now its growth is less than that of solar and wind. In order for geothermal energy to expand its market share several situations need to occur. These situations involve technology, location, transmission, construction, and risk.

New technology to advance geothermal energy has already been discovered. However, it is still in a development and implementation phase. In order for geothermal power to increase, the EGS and binary technology need to be perfected so that they reach an efficiency that makes it economically viable to invest in geothermal energy on a wide scale.

The main limitation for geothermal energy is location. If EGS and binary technology can be improved then this will allow for an extremely large range of potential geothermal areas to become resources for geothermal power plants. Right now these areas are either not hot enough or do not have enough of a flow rate to be used in a power plant. Areas that have less geothermal activity could become viable if the

---

[180] "GEOTHERMAL ENERGY - Worldwide." Geothermal Education Office, Accessed March 1, 2015. http://geothermal.marin.org/geomap_1.html.
[181] Roberts, Billy. "Geothermal Resource of the United States." 2009. Accessed March 1, 2015. http://www.nrel.gov/gis/images/geothermal_resource2009-final.jpg.

technology for drilling wells, analysis, and management of underground brine reservoirs could be improved to function at deeper levels.

Another limitation is the transmission for direct-use geothermal power. Right now direct-use geothermal power relies on these transmission pipelines to transport the hot fluids to buildings and other sites for heating. For this process to greatly expand the transmission lines would need to have increased insulation ability so that the fluid can travel farther without losing too much heat. Even still the actual construction of a network of transmission lines would be very expensive.

Relative to wind and solar power, geothermal power plants take four to eight years longer to complete.[182] This is due to the extreme effort involved in managing the risks of constructing a geothermal plant. New advanced technologies for sufficiently managing and solving these risks have yet to be developed. The current technology requires a lot of time, effort, and money and is a contributor to geothermal energy's high initial cost. These risks involve pollution from the brine and working fluids and the risk of subterranean alterations caused by the geothermal plant that can negatively alter the landscape.

Conclusion

In conclusion, geothermal energy is a viable renewable resource that has many advantages. In order to fully understand it one should be aware of its processes, methods of use, relationship with the economy, place in the environment, how it is affected by social factors, and what can be done to increase its market share.

Problems

1. A geothermal plant receives brine from a production well with a flow volume of 70 m^3/s and a volumetric mass of 1 g/cm^3 at 300°C and returns the brine to an injection well at 40°C. The plant has a heat delivery of 1,000 MJ/h. What is the heat capacity in J/(kg*K)?
2. A geothermal plant with a heat delivery of 2,000 MJ/h receives brine from a production well with a volumetric mass of 1 g/cm^3

[182] "U.S. Has Large Geothermal Resources, but Recent Growth Is Slower than Wind or Solar." U.S. Energy Information Administration - EIA - Independent Statistics and Analysis, November 18, 2011. Accessed March 1, 2015. http://www.eia.gov/todayinenergy/detail.cfm?id=3970.

at 250°C and returns the brine to an injection well at 40°C. The brine has a heat capacity of 0.005 kJ/(kg*K). What is the flow volume in m³/h?

3. A geothermal plant with a heat delivery of 3,000 MJ/h receives brine from a production well with a flow volume of 45 m³/s and a volumetric mass of 1 g/cm³ and returns the brine to an injection well at 40°C. The brine has a heat capacity of 0.005 kJ/(kg*K). What temperature (in kelvins) is the incoming brine?

4. What is the temperature of the brine at a location with a geothermal gradient of 0.025 °C/m and a ground temperature of 662 °F where a 400 m well has been drilled?

5. How deep of a well is required to obtain a resource temperature of 110 °C. The location has a geothermal gradient of 25 °C/km and a ground temperature of 212 °F.

Solutions
1. 0.916 J/(kg*K)
2. 1,907.94 m³/h
3. 536.37 K
4. 360 °C (680 °F)
5. 400 m

Index

www.ingramcontent.com/pod-product-compliance
Lightning Source LLC
Chambersburg PA
CBHW032014170526
45157CB00002B/692